Keeping a Cow

Family cow

Sunnyview Little Princess 30th. A famous Illawarra cow that holds the Australian lifetime record for butterfat production and almost broke the world record.

Keeping a Cow

Jim Wilson

Third Edition

Kangaroo Press

Preface

As a result of a conversation early in 1977 between myself and Dr John Drinan, former Senior Lecturer in Animal Husbandry at Hawkesbury Agricultural College (now the University of Western Sydney, Hawkesbury), I was to plan and run a weekend school entitled 'Keeping a House Cow'. Fifty-three people attended and a booklet was printed to cover the main topics. These schools were repeated on an annual basis for a number of years and seemed to be meeting a real need.

My original intention for the book was to provide the sort of information relevant to many thousands of Australians who live on the outer fringes of our cities and towns, those who own a hobby farm or those who live or plan to live alternatively or engage in subsistence level agriculture. Demand for the book however has come from a much wider clientele hence a second edition in 1987 and the third in 1990. *Keeping a Cow* has also been found helpful by some commercial dairy farmers and increasingly it has found its way into high schools. I have been told the main reason for the success of the book (as *the* book on the subject) is that it has been deemed a landmark publication in that it reached a totally new level of integration of theory and practice for a book of its type. It is my hope that this updated version will continue to prove a useful resource in the understanding and care of animals and their utilisation for the benefit of mankind.

Acknowledgments

The House Cow Schools referred to were all run in conjunction with the New South Wales Department of Agriculture. Messrs Reg Smith, Tony Dowman, Hugh Allen, Ross Watson and Miss Jayce Morgan (all Department of Agriculture Officers and ex-students of Hawkesbury) and Messrs Ken Hide, John Drinan and Norm Jones (Hawkesbury staff) all contributed to the success of these schools and therefore indirectly to the publication of this book.

I am grateful to the following for permission to use illustrations: Babson Bros. Co., builders of SURGE, NSW Department of Agriculture, and Alfa-Laval Pty Ltd (formerly the De Laval Separator Company) for illustrations from the De Laval Handbook of Milking, 1963.

About the Author

Jim Wilson has owned dairy cattle for almost 30 years and has lectured in dairying for much of that time. He also owns and manages a 450 hectare cattle and timber property near Dorrigo in New South Wales. His current position is Lecturer, Faculty of Agriculture and Rural Development, University of Western Sydney, Hawkesbury.

Under supervision children can do an excellent job rearing calves.

Third edition published in 1990
Second edition published in 1987
First published in 1983 by Kangaroo Press Pty Ltd
3 Whitehall Road (P.O. Box 75) Kenthurst 2156
Printed in Hong Kong through Colorcraft Ltd

ISBN 0 86417 313 X

Contents

This cow literally 'fell off the back of a truck' and subsequently walked with a slight limp. She was no use in a commercial herd, yet proved an ideal house cow. She couldn't jump fences or run away, had nice big perpendicular teats and a well attached udder which was not too far above the bucket.

Richmond Honour, a famous Jersey cow that has held a state record for milk production for fifty years.

An attractive Guernsey cow.

Introduction

If you like animals, have sufficient land (see p 49 Feeding) and live a reasonably settled domestic existence, keeping a cow is both pleasant and profitable. In fact, on a time, effort and return-on-invested-capital basis, you are more likely to come out in front with a cow than with homegrown poultry or fruit and vegetables. However, this is not a book about economics of the production of unadulterated food, rather it is my hope that it will help house cow owners and potential owners to make wise and timely decisions, to be aware of possible problems and to gain confidence in cow husbandry.

It has also been the author's aim that this book may prove a useful reference for the stud cattle hobbyist, Rural Youth member or student interested in keeping or exhibiting cattle, or entering a judging competition.

In Australia keeping a cow may not necessarily involve you in a great deal of expense. Cows are not expensive and it is sometimes possible to borrow one from some hobby breeder of stud dairy cattle providing you arrange A.I. to the right bull and forgo the calf. A good house cow may cost up to $500 but the average family may easily spend $1000 a year on dairy products. As hand feeding and housing requirements are minimal in Australia the economies are there, providing any irrigation scheme is wisely used and does not involve a great deal of capital expense or use of excess town water.

At times throughout the book various matters are dealt with which pertain to cow health. Whilst some may feel confident to attempt to treat the cow themselves, let me say at the outset that it is *always* wise to consider calling a vet if you feel out of your depth. The pooled wisdom of the family, experienced friends and neighbours usually helps the decision making process.

Some of the material in the following chapters may appear unnecessarily scientific and detailed. It has been included, however, for those interested in 'why' as well as 'how'. Farming and husbandry are about problem solving and decision making. Sometimes the theoretical helps light dawn and helps one gain confidence in making decisions.

1 Selecting a House Cow

The *ideal* cow is quiet, her teats are well placed, perpendicular, of optimum and even size and her udder is well attached. She has a strong constitution and produces large quantities of milk and goes in calf easily.

Such a cow might also cost more than the average house cow owner wishes to pay. However, as the criteria for a house cow and a commercial dairy cow are not identical it is possible to milk a 'bargain' providing we are not aiming to win first prize in the local show.

Where to get a cow

Cows culled from commercial dairies for low production, injury or because they only have three functional teats may be obtained cheaply and prove quite satisfactory house cows.

It is sometimes suitable to buy a cow with chronic mastitis providing she has it in only one quarter (see p. 62 section on mastitis).

In dairyfarming areas you usually get what you pay for and, as a rule, suitable cows will be cheaper than on the outskirts of metropolitan areas where such cows are not so abundant. It is possible to pay more than market value often through ignorance of the seller as well as the buyer.

Dairy farmers, milk tanker drivers, AI technicians, officers of the Department of Agriculture, dairy cattle breed societies, vets and livestock carriers may all be able to put you onto the track of a suitable cow. Livestock columns in rural and local newspapers provide an excellent medium for potential buyers and sellers to come together. People of long standing in the district who own cows often know who to see or where to go.

Remember, if you feel an amateur at the stock buying game, try to get someone who knows cows to go with you. Taking other members of the family may also be a good idea as all should feel some sense of responsibility and ownership.

Local auctions where anything is sold are not usually the best places to buy a cow. However, auctions run by reputable agents, often exclusively for cattle, are another matter. The problem at an auction where you are bidding against regular buyers is that the amateur can easily become bewildered. An experienced friend or the agent himself may bid for you. Agents will also arrange suitable cows on an individual transaction basis but the price is usually higher as the cow does come with a sort of guarantee of being sound because the reputation of the agent is involved.

Buying calves and rearing them is an excellent, though long term, approach to cow ownership.

However if you are not really sure about the whole idea, one possibility is to ask someone who has a cow if you could share management, milking and feed costs for a while, as most cows give too much milk for one family anyway. In this way you can gain confidence and determine whether cow ownership is for you.

Another idea is to try looking after a cow while the owner is on holidays. In fact, it is a very good idea to consider dual ownership or at least dual management of a house cow with a neighbour.

A properly managed cow should provide enough milk for more than two families for most of her lactation. Shared milking means you aren't tied down day in day out. Of course you can share your cow with a calf if the cow has plenty of milk and you can't always be there to milk her.

What to look for in a cow

The main criteria in a house cow are a quiet temperament, teats that make for easy hand milking, an udder which gives sufficient daily milk and persists in doing so right up to drying off time,

and a problem-free reproductive system which enables the cow to go in calf and calve easily.

Temperament is not hard to assess. Prior to purchase, approach the cow, try patting her on the head and neck then gently run your hand down her flank to her udder and then try squeezing her teats to see if she is easy to milk.

If you or your more experienced friend are still in one piece you are at least at first base. If she repeatedly kicks viciously when you do this slowly and gently forget her. It is just not worth bothering with a cranky cow.

However, some cows may kick just in a fidgety way, perhaps because their udders are a bit sore. A legrope may be necessary with these.

A number of factors are involved in temperament and an initially nervous or flighty cow can with gentle and firm handling, prove satisfactory. Cows may be upset simply because they are isolated from their herdmates. I have known otherwise reasonable cows to become quite unmanageable and cease lactating when cut off from other cattle. Check out these circumstances. Note, therefore, that flightiness and nervousness are not the same thing as fidgetiness or crankiness. Cows with the first three problems can often be won over and settle down well in time; lack of human contact or sometimes even cruelty or insensitive handling has been the cause. Real crankiness and bad temperament is, however, inherited and is hard to live with and won't be greatly improved.

One thing to be on the watch for is that sometimes at auctions cows (and horses) of intractable temperament will appear passive and submissive because they have been given tranquilizers.

When purchasing a cow try to ascertain what the cow is like with her usual owner—he may give you a demonstration.

If you have doubts it may be possible to arrange some sort of guarantee.

A freshly calved cow, or one close to calving is the best guarantee against reproductive problems and you can try out hand milking the lactating cow. Feel the udder for lumpiness as this is a sign of mastitis damage.

Do not try to milk a dry cow or one prior to calving. To break the natural seal in the teat is to invite mastitis infection.

If the cow is a very "tight" milker and you have to work very hard, or the teats are very short, steer clear of her unless you have a milking machine.

As a rule a cow over 10 years is not such a good investment unless you get her cheap. She may still look great but is usually more prone to mastitis and is harder to get in calf than a younger animal.

However, as a first time round experiment, she

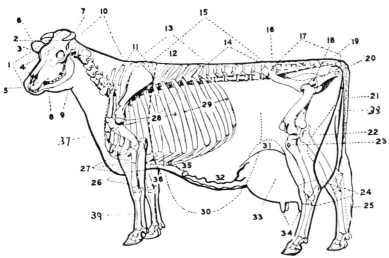

1. head	11. shoulder	21. tail	31. flank
2. forehead	12. crops	22. thigh	32. mammary or
3. eye	13. chine	23. stifle joint	milk veins
4. face	14. loin	24. hock	33. udder
5. muzzle	15. back	25. switch	34. teats
6. ear	16. hip	26. brisket	35. milk well
7. horn	17. rump	27. chest	36. heart girth
8. jaw	18. thurl	28. foreribs	37. dewlap
9. throat	19. tail head	29. back ribs	38. escutcheon
0. neck	20. pin bone	30. barrel	39. cannon bone

could be the sort to start with as she is very likely to be quiet. You can get some idea of a cow's age by grey hairs, a deeper body and udder, and the wear on her teeth (if she has none she's sure to be old!).

Can you tell a good producer by looking at her?

Yes, an experienced person can, on average, select better milkers on their looks, all things being equal, but knowledge of a cow's production, her ancestors and offspring all increase the accuracy of assessment.

Some experienced dairymen, judges and breed classifiers can pick, at a glance, high producing dairy cows with almost monotonous regularity.

A cow may be likened to a motor car. The motor is the inner driving force and the chassis and body work are the containers that hold the driving parts together and keep the whole show on the road. It is no use a cow having a powerful motor or milk producing ability if her udder gives way while she is still a young cow, or her feet and legs give in and she cannot get about. Likewise, a cow can have many attractive features from an aesthetic point of view (attractive bodywork) but still be useless because she has little drive to produce milk. Two well-known sayings in the dairy industry are 'no udder, no cow' and 'its no use having an udder if you haven't got a cow to fill it'. These two views held together and in balance hit the nail on the head.

Beyond the essentials of high performance motor and strong frame and functional body, added features simply enhance the beauty of the cow as they do the car—optional extras that are relatively unimportant to the main purpose, but giving pleasure to the owner. Included here are 'gun-barrel' toplines, attractive colour markings and level and tidy rumps.

Breeds of dairy cattle

In Australia a number of dairy breeds are found in their pure state. Listed roughly in order of increasing size and decreasing milk solids content they are: Jersey, Australian Milking Zebu, Guernsey, Ayrshire, Illawarra (formerly Australian Illawarra Shorthorn-A.I.S.), and Holstein-Friesian.

Dual-purpose cattle such as Red Poll, Milking Shorthorn and Simmental are also milked on.

Jerseys are, by and large, the most suitable cows.

They eat less, are easier to lead about and their udders are not too high above the bucket. Their cream also rises more quickly and is easily skimmed off. Friesians are only suitable if you share them with one or two calves. Otherwise you could spend a lot of time hand milking only to "drown" in milk. Illawarras have also been popular at times as they cross well with beef breeds and their body fat is white rather than the yellow of the Channel Island breeds.

Some house cow owners may wish to register their animals with a breed society and make a hobby out of raising some registered calves. This can be real fun as a family enterprise, especially if, under supervision, the children rear the calves. They may be able to include this experience as credit for a Guide or Scout badge, or Rural Youth or school project. The late primary and junior high school years, when young people can run into problems are just the years when they can get really motivated about animals. What better time to inculcate some solid rural values? Calf rearing competitions and dairy cattle judging competitions are also gaining in popularity in some areas.

One last word on selecting your cow: if you get one that does not suit, do not be a martyr—ship her off and get another.

Why not a dairy goat?

On a nutritional basis the milks are similar. A small fraction of the human population is, however, allergic to cows milk (some I have been told are only allergic to *pasteurised* cows milk), or do not digest it readily. Goats of course, produce much less milk than cows and also tend to be seasonal breeders, kidding in spring. They are often more expensive to buy (proportionately) than cows and are fussier feeders and more prone to chronic infestation with internal parasites. On the positive side they are probably more intelligent and sensitive, enjoy human company more and make better pets, hence they do fret more.

Some of the material in this chapter was taken from an article by the author which appeared in *Hobby Farmer*.

A prize winning dairy cow, Meadowhaven Jenny 8
The Dairyman

A prize winning Hereford cow, Tiverton Corisande 40
Stephens Livestock Photography

Form is related to function. This comparison between a dairy and beef cow certainly makes the point. The dairy cow may be eating twice as much as the beef cow but she turns it into milk rather than muscle and fat. A good milker is angular, with a large barrel and capacious udder.

2 Cattle Behaviour

Behaviour in cattle is like that of other animals, having a genetic, physiological and psychological basis. Much of it is instinctive and is displayed even by animals raised in complete isolation. However, cattle are social creatures and many of their behaviour patterns are modified by continued interaction with the environment.

The communication gap between man and animals still remains immense. However, anyone who has owned a pet — particularly a dog, or done a lot of work with horses, knows that it is not only possible to communicate our wishes to animals, but also for them to tell us things. In fact they are like children and will 'put it all over us' if we are not prepared to assert ourselves. In other words there will be a tussle to see who is boss.

By knowing something of cattle behaviour a house cow owner may not only be able to quieten his cow and do more with her, but also train her in such a way that his labour is reduced. Also, if the animal is hungry, thirsty, distressed or sick or on heat, someone with the knowledge of cattle behaviour will pick up any message the animal is communicating even though it may be by very subtle cues rather than an overt display. Cattle also communicate aggression by means of threat postures and intention movements. Reading these signs correctly can save one from serious injury.

Systems of behaviour

Behaviour systems include ingestive, eliminative, sexual, care giving, care soliciting, agonistic (combative), shelter seeking, investigatory, and, if you like, abnormal or maladaptive.

A system of behaviour can be broken up into a number of related patterns, and these patterns further categorised into a number of sequences which are usually but not necessarily, strung together.

Ingestive behaviour for example consists of how a cow grazes using her tongue etc., how often, for how long a period, what times during the day, when and how she drinks, ruminates, and so on.

A heavy milking cow grazes and ruminates about 16 hours out of 24 hours and cows seldom sleep.

A dog can eat its daily requirements in 5 minutes and sleeps for much of the time. Some carnivores sleep 22 hours in 24.

A cow's ingestive behaviour, as with her eliminative behaviour, is closely allied with her anatomy and physiology. The same can be said of her sexual behaviour.

However, some aspects of cattle behaviour are what is known as species specific. Chin resting and mounting in cattle sexual behaviour is an example. Another interesting example of species specific behaviour is in the way cows relate to their calves.

Cattle are a 'lying out' species and leave their young hidden or curled up near a fence while they go off to graze or drink, whereas sheep are a 'follower' species and lambs keep up with their mother at all times ('everywhere that Mary went the lamb was sure to go'). So if your cow has a calf which 'disappears' do not get in too much of a panic as it is probably around somewhere.

Cows are essentially very *social* creatures. They have a well developed social hierarchy which varies from simple to complex. When herds are very large this social order tends to break down in much the same way as human society does in very large cities. In large herds cows are no longer able to recognise all other members as individuals and cannot recall 'who fought who and who won'.

A cow's position in the social hierarchy is determined by whether she wins in displays of aggressive interaction against her herd mates. These interactions range from violent body contact to very subtle threats and cues. The 'peck order' is such that while cows may become very tolerant of their herd-mates in close proximity, by and large each cow has an area round her she does not like being violated. If it is violated, she tries to move away if the other cow is dominant to her, or she threatens or attacks if the other cow is subordinate and does not quickly retire. Age, size and length of time in the herd all tend to be positively correlated with a cow's position in the 'peck order'.

Aspects of cattle behaviour and the house cow owner

Socialisation to humans If social species such as cattle are introduced to human beings very early in life then humans will be included among the species with whom they will relate socially. Bond formation between a calf and its mother occurs most rapidly and completely if the cow is allowed to clean and suckle the calf. Movement and smell are the two most important things that attract the cow to the calf. Calves, like young of many species, imprint onto the first large animal around —mostly their mother. However, it is not difficult to get a calf to imprint to humans as a substitute mother *during the first few days of life*. Feed reinforcement being a powerful tool. However, beyond the first few days it becomes increasingly difficult to establish this close attachment and the bond formed will never be as complete.

If dairy heifers are not socialised to humans early in life then it will be difficult for them to accept humans as a substitute calf later on when we try to milk them. Hence training for a quiet temperament starts at birth. The reason we can manage cattle is that we plug into their behaviour patterns — into their world. It is possible for cows to become quite affectionate towards their keepers and lean their neck onto your body and seek to be scratched and patted. One commercial dairyman not far from where the author lives calls each of his cows by name when he wants them to enter the milking bail.

Training Cows are very much creatures of habit and they will take a while to get used to any change in routine. Reinforcement will lubricate any change. Feed is often used as a positive reinforcement, a scratch behind the ear will also break down apprehension.

However, we cannot train an animal to perform some task if it is outside of the range of its instinctive behaviour pattern. For examle, cattle have no elaborate patterns of eliminative behaviour such as a dog or cat. If you belt a cow every time she messes in the bails she could well get worse instead of better. It is not very easy to toilet train a cow.

The use of positive and negative rewards (reinforcement) They say donkeys are trained with 'the stick and the carrot' — this is a generalisation. Some cows will require more of one than the other. Experience will enable you to tell whether a cow is difficult to handle because she is 'acting up' and being fussy or fidgety, or whether she is actually nervous, lacking confidence or upset. One good thump in the lower ribs will often settle down the

"Stop tapping your foot."

Hoards Dairyman

bossy, confident or fidgety cow but quiet and gentle reassurance is the sort of handling the nervous or timid cow needs.

Remember, cows are a social species and are used to being dominated. We must, therefore, strive to have the cow fully socialised to, and trustful of humans but as the submissive partner in the relationship.

If you feel you have given your cow every opportunity to be pleasant and co-operative and she continues to be the opposite, I would suggest you trade her in on a new model. Why waste time fiddling about, or risking injury every day of your life? If you are one of those 'till death do us part types where animals are concerned then seriously consider planning her demise.

Safe handling of cattle

Cattle are big, heavy and strong and there is always a risk of injury or even death when handling them. However, this risk can be greatly reduced if they are handled correctly, and we do not keep calves from temperamental (wild and cranky) cows as temperament is inherited.

To handle animals correctly we need to know something of how they behave and then develop techniques and facilities which are atuned to their built-in behaviour patterns.

Cattle behaviour and cattle handling

Being social creatures, cattle naturally stick

gether. They develop a defined social dominance order and timid animals keep a reasonable distance from the more aggressive ones.

This idea of 'reasonable distance' is even more important in the relationship between species, particularly if one is a herbivore and the other is a carnivore and possible predator. This distance is known as *'flight distance'* and is common to most wild animals. As soon as one species which may pose a threat moves closer than a certain distance the threatened species moves off. However, if the threatened animal cannot get away or is surprised by a threatening species at close range it may turn and attack. Hence a cow may try to run away or to lure you away from her calf, but if you get too close she may attack you. Bulls likewise sometimes only attack those who get closer to them than a certain distance — often about 2 to 3 metres.

Flight distance is one of the key behavioural patterns utilised in cattle handling and moving and yet it must be overcome or greatly reduced for the purpose of practices such as milking.

One of the key factors in reducing flight distance is *early socialisation* to humans and this is achieved by human contact with calves from day old onwards (usually artificially reared). Other important elements are selection for quiet temperament and kind, quiet, but firm handling throughout life, and hand feeding. But early socialisation during the first few days (known as the critical period) will often achieve more than months of hand feeding and handling later.

Cattle react quickly, but think slowly. This is important to remember when calling cattle to come to feed or a new paddock. One call is often not enough and four or five may be necessary before the message gets into their skulls. The fact that cattle react quickly but think slowly is also important when trying to get them through a gate which is normally kept shut. If you press them too closely and invade their flight distance they will very likely break away. If this situation has occurred and the cow(s) are refusing to comply it is usually wise to back off and just contain the cattle while they settle down, see the open gate and finally wander through.

A further aspect of cattle behaviour crucial to successful handling is the desire of cattle to stay together as a mob, particularly if threatened. If one or more animals are separated from the rest they will automatically try to get back to the remainder of the herd — a sort of 'magnetic return'. In a round yard cattle tend to wheel in an anti-clockwise direction.

A final aspect of cattle behaviour significant to cattle handling is known as the *'point of balance'* (see top photograph, page 67) just behind the shoulder. If you approach an animal in front of that point it will tend to back off, and if approached from behind it will go forward to maintain flight distance (see page 67). Hence, approach should be made and maintained at right angles to that point if you wish to get close to the animal. A quiet and gentle approach obviously reduces flight distance and boisterous behaviour and sudden noise or movement increases it.

Investigatory behaviour (eyes alert, nostrils quivering, animal advancing slowly and tentatively, sniffing in short puffs and tongue licking once nose touches) is a primary factor in all patterns of behaviour and can be utilised in cattle handling by giving a cow time to investigate a new situation rather than rushing her. A trail of hay and a little patience may get a cow through a gate better than all the yelling and hitting.

In the above photos the cow investigates by looking and approaching then sniffing and finally licking.

Cutting out, moving and restraining cattle — 'cow sense'

Know how animals respond. Having selected for quiet temperament in the breeding program (cull cranky animals), and raised them correctly, the next step is to handle and train them. Key factors are:

1. Quietness and firmness. Talk quietly. You have to win their confidence.

2. Consistency — cattle are creatures of habit.

3. Right placement of body, use of cattle cane or length of bamboo to extend ability to block and cut out.

4. Ability to anticipate what animals will do; know when to move slowly or quickly.

5. Know when an animal is in a 'threatening' posture.

6. Learn to work in teams and change body placement and *tactics* instantly. Strategies remain the same.

7. Design and locate gates, fencing yards, crushes, etc. to fit in with the way stock move and behave.

8. Knowing when to back off and let the animal relax and gain its bearings.

Movement

Moving a whole mob of cattle is easier than cutting out one or more and trying to move them away from the rest. It is this operation of separating animals that want to stay together, whether carried out in a paddock, yard or lane, that requires skill and perseverance and carries certain risks to personal safety. Handling animals in the race, crush or in the case of a dairy cow, the milking operation, transporting animals or leading with a halter may also prove occasions where the operator runs the risk of causing injury to himself or the stock.

When cutting out cattle, use fences to work cattle toward the front or rear of mob and then step in and separate. At this point determined effort and some shouting etc. may be required to prevent the animal(s) rejoining the mob. They will try to either turn back along the fence or circle out behind the operator and back to the mob. This is where anticipation and teamwork are vital to success. Once the animals have been cut out, keep them moving and have the gate(s) placed in such a way as to aid and not hinder the job in hand, i.e. anticipate your gate position ahead of time.

The principal means whereby cattle defend themselves are by charging and butting (head) and kicking. Due respect, caution and ability to anticipate and react quickly will save you from many an injury.

Readers who have only one animal may wonder at the relevance of this chapter to their situation. The chances are, however, that your cow will end up in someone else's paddock one day — or their stock in yours.

Cattle, like other domestic animals, were created for man to use and we can mostly get them to do what we want them to do if we go about it the right way.

Restraint

Options and Approach

While brute force may be necessary in restraining animals, a knowledge of how animals react and the correct position of the operator can save many an injury.

Quiet and experienced handling sometimes reduces or eliminates the necessity for heavy restraint. With animals you do not know or temperamental ones always run your hand down from the top of the animal when wishing to handle the underline, legs, udder, etc. and keep a hand placed on the animal. Our body reacts much more quickly to *feel* than to *sight*. It is possible to feel through your fingers when the cow tenses her muscles as she prepares to kick, well before your eyes see the hoof.

Equipment and Methods

1. Fixed facilities: such as crushes, bail heads, races

2. Equipment: such as halters, nose grips, nose rings, bull staffs.

3. Use of pressure points e.g. bending up tail, rope around flank, holding tongue under jaw, tail under flank and around back leg, etc.

Bending a cow's tail up will often help to keep her still (providing her head is, of course, securely restrained).

If a cow lies down and will not get up, pour water down her ear or smother her by pressing the heel of each hand in a nostril and grip the side of the mouth with the fingers.

In summary, in handling cattle, plug into the cow's world.

We can successfully *raise* calves because we plug into the mother slot and become the calves' substitute or surrogate mother. Like-wise we can *milk* cows because we are accepted by the cow as a substitute calf. We *separate (cut-out) and block* the progress of animals by utilising the principle of flight distance. However, we have to balance this technique with that of taming whereby we can *approach*, handle and catch animals if necessary.

When we dominate a cow and get her to do what we want we are assuming a position of a more dominant member of the herd. That is why cows will 'try us out' and see if they can win in the 'battle of wits' and on occasions, a battle of brute force, to see who will be the more dominant partner in the relationship.

By using mechanical advantage in tried methods of restraint and utilising the cow's behaviour patterns, we can even win in the battle of brute force.

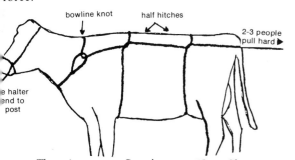

Throwing a cow. See also page 70 on Showing.

A suitable head bail   1. Open   2. Closed.

Holding a cow's mouth open. At times it is necessary to administer pills, boluses, gastro magnets, etc. or to inspect inside a cow's mouth. One method of holding the mouth open is illustrated. Grasp the tongue with the left hand and pass it under the chin to the right hand. The cow must be restrained in a head bail. Be very careful when placing your hand inside a cow's mouth as the tongue will endeavour to push whatever is in her mouth to one side or the other and chew it.

When drenching a cow, stand in the same position as in the picture above, lift the cow's head (you may need the aid of a halter or nose grips) and by placing the fingers of your right hand over the nose and into the right side of the mouth, lift up. As the cow opens her mouth place the drenching gun or bottle into the left side of the mouth in the space between the incisors and the molars.

A moderately priced type of nose grip shown in both the open and closed position.

3 Cattle Reproduction

Reproduction, udder development and milk secretion are controlled by complex hormone systems which are linked closely together, and therefore pregnancy and parturition are essential to initiate milk production. It is possible to artificially induce lactation in empty cows or heifers with the use of hormones, but research and experiment to date have not led to a fundamental review of cattle reproduction although there have been successes in bringing barren heifers into commercially viable lactation. The hormones used also sometimes trigger normal cycling in previously barren animals. However, artificially induced lactations are not generally as productive as those following the birth of a calf and many of these lactations fail completely.

Whether solutions will be found to this failure rate, we are yet to see, but for the present a cow needs to have a calf in order for us to get milk out of her. Once she has had a calf, however, she may milk on for years as a single lactation. Some cows are genetically more persistent producers over time than others which may dry off six to nine months after calving no matter what.

Animals from which the ovaries have been removed (speying) are usually the surest bet for persistency. Empty (non-pregnant) cows also have less tendency to drying off.

However, under normal circumstances, it is most efficient to calve cows about every twelve months. This annual cycle not only results in the most milk, but the accompanying calf may act as a possible future replacement for the cow, or be sold for profit. Annual calving also allows the possibility of the speediest increase in genetic potential. If the cows are mated to the best bulls in artificial insemination centres the chances are that the calves will be better all round milkers than their mothers.

The cow's hormonal physiology

The whole of a cow's life and usefulness is tied up with her hormonal regime. The most important hormone-producing gland is the pituitary (hypothalamus) situated at the base of the brain.

The hormones it produces sometimes trigger other glands such as the ovaries, adrenals and thyroid to produce hormones also. All these hormones circulate in the blood stream and their varying concentrations may trigger feedback mechanisms which regulate their production or, at other times, once a certain concentration of a hormone is reached, it may stimulate the release of some other hormone. For example, the blood level of progesterone increases during pregnancy, until the point is reached when its concentration in the blood towards the end of gestation triggers the production by the ovary of increasing amounts of oestrogen which in due course triggers a decrease in circulating progesterone and then around parturition a vast increase in the concentration of the hormone prolactin which helps initiate lactation.

The pituitary gland works like the playing conductor of a small orchestra, the musicians being the glands and the hormones they produce being the type of sound produced by their particular instrument. Some musicians (glands such as the pituitary and ovaries) play more than one instrument. Each musician has its musical score (genetic script), but *they also listen to each other* and at various stages in the music are triggered to play (produce hormones) by the sound of the melody as it flows, their own musical score, and also a trigger from the conductor (pituitary gland). At times such as around parturition (calving) the music reaches a crescendo.

The oestrus cycle

This is a simple example of the ebb and flow of hormones in the cow and it is very important that the house cow owner has the ability to recognise when a cow is on heat (in season or in oestrus) as a cow is physiologically ready for conception only during oestrus or heat period. Thus occurs every 21 days (ranges from 18 to 22 days) in non-pregnant cows, and lasts from 15 to 18 hours. It is only during this period that a cow will stand for a bull. At the onset of oestrus the blood levels of oestrogenic hormones are high and level of progestins low.

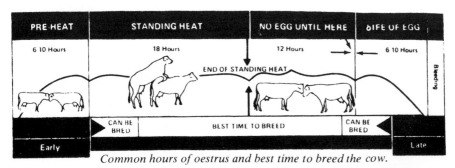

Common hours of oestrus and best time to breed the cow.

How important is heat detection?

Unless there is a bull about the need for effective 'heat detection' cannot be over emphasised — NO OESTRUS — NO INSEMINATION — NO CALVING.

Oestrus behaviour

Cows being a social species are continually interacting. These interactions heighten in intensity when an animal is in season. They attract from others and display themselves, not only mounting, sniffing and chin resting behaviour (all fairly specific to reproduction) but also agonistic (fighting) and bellowing behaviour. They will sometimes wander off in search of a bull and have been known to swim substantial rivers. Some cows are very demonstrative when in season; at the other end of the spectrum are those that are very hard to pick (sometimes known as 'silent' heats or 'shy' breeders).

A cow on its own obviously has difficulty in displaying many facets of oestrus behaviour. However bellowing, being agitated and temperamental, failing to let down milk, walking around the perimeter of a paddock with an agitated look, or clearing right out and winding up with some bull a kilometre away are all quite likely to occur. Sometimes a cow will even try to mount other species including humans.

The darker cow is on heat. Investigative sniffing is followed by chin resting and then mounting. If she is not quite on heat or just going off she will allow chin resting but will move away when the other cow mounts.

Flehman Response; The cow's nose is characteristically turned up. This is sometimes associated in cows and bulls with sexual investigation. The light coloured cow has just urinated.

Mounting behaviour. The cow underneath is in season. A cow in season also usually attempts to mount other cows in season like herself.

17

Other signs of oestrus

Enlargement of the vulva from which there may be a clear mucous discharge and mucous on the tail. Hair ruffled at the base of the tail and down either side of the barrel may indicate that the animal has been mounted recently.

Blood on the tail head is usually an indication that a cow has just gone off heat. A small quantity of blood is often released by the ovary at ovulation which occurs at the end of, or just after oestrus. So if you see a little blood and the cow does not appear to be on heat, mark off the date twenty days hence on the calendar and watch her closely at that time.

Gestation chart for all classes of stock

Date of Service	Expected Date of Parturition						
	Mares (340 days)*	Cows (283 days)*	Ewes (150 days)*	Does (147 days)*	Sows (116 days)*	Bitches (63 days)*	Rabbit Does (31 days)*
1 Jan.	7 Dec.	11 Oct.	31 May	28 May	27 April	5 Mar.	1 Feb.
8 ,,	14 ,,	18 ,,	7 June	4 June	4 May	12 ,,	8 ,,
15 ,,	21 ,,	25 ,,	14 ,,	11 ,,	11 ,,	19 ,,	15 ,,
22 ,,	28 ,,	1 Nov.	21 ,,	18 ,,	18 ,,	26 ,,	22 ,,
29 ,,	4 Jan.	8 ,,	28 ,,	25 ,,	25 ,,	2 April	1 Mar.
5 Feb.	11 ,,	15 ,,	5 July	2 July	1 June	9 ,,	8 ,,
12 ,,	18 ,,	22 ,,	12 ,,	9 ,,	8 ,,	16 ,,	15 ,,
19 ,,	25 ,,	29 ,,	19 ,,	16 ,,	15 ,,	23 ,,	22 ,,
26 ,,	1 Feb.	6 Dec.	26 ,,	23 ,,	22 ,,	30 ,,	29 ,,
5 Mar.	8 ,,	13 ,,	2 Aug.	30 ,,	29 ,,	7 May	5 April
12 ,,	15 ,,	20 ,,	9 ,,	6 Aug.	6 July	14 ,,	12 ,,
19 ,,	22 ,,	27 ,,	16 ,,	13 ,,	13 ,,	21 ,,	19 ,,
26 ,,	1 Mar.	3 Jan.	23 ,,	20 ,,	20 ,,	28 ,,	26 ,,
2 April	8 ,,	10 ,,	30 ,,	27 ,,	27 ,,	4 June	3 May
9 ,,	15 ,,	17 ,,	6 Sept.	3 Sept.	3 Aug.	11 ,,	10 ,,
16 ,,	22 ,,	24 ,,	13 ,,	10 ,,	10 ,,	18 ,,	17 ,,
23 ,,	29 ,,	31 ,,	20 ,,	17 ,,	17 ,,	25 ,,	24 ,,
30 ,,	5 April	7 Feb.	27 ,,	24 ,,	24 ,,	2 July	31 ,,
7 May	12 ,,	14 ,,	4 Oct.	1 Oct.	31 ,,	9 ,,	7 June
14 ,,	19 ,,	21 ,,	11 ,,	8 ,,	7 Sept.	16 ,,	14 ,,
21 ,,	26 ,,	28 ,,	18 ,,	15 ,,	14 ,,	23 ,,	21 ,,
28 ,,	3 May	7 Mar.	25 ,,	22 ,,	21 ,,	30 ,,	28 ,,
4 June	10 ,,	14 ,,	1 Nov.	29 ,,	28 ,,	6 Aug.	5 July
11 ,,	17 ,,	21 ,,	8 ,,	5 Nov.	5 Oct.	13 ,,	12 ,,
18 ,,	24 ,,	28 ,,	15 ,,	12 ,,	12 ,,	20 ,,	19 ,,
25 ,,	31 ,,	4 April	22 ,,	19 ,,	19 ,,	27 ,,	26 ,,
2 July	7 June	11 ,,	29 ,,	26 ,,	26 ,,	3 Sept.	2 Aug.
9 ,,	14 ,,	18 ,,	6 Dec.	3 Dec.	2 Nov.	10 ,,	9 ,,
16 ,,	21 ,,	25 ,,	13 ,,	10 ,,	9 ,,	17 ,,	16 ,,
23 ,,	28 ,,	2 May	20 ,,	17 ,,	16 ,,	24 ,,	23 ,,
30 ,,	5 July	9 ,,	27 ,,	24 ,,	23 ,,	1 Oct.	30 ,,
6 Aug.	12 ,,	16 ,,	3 Jan.	31 ,,	30 ,,	8 ,,	6 Sept.
13 ,,	19 ,,	23 ,,	10 ,,	7 Jan.	7 Dec.	15 ,,	13 ,,
20 ,,	26 ,,	30 ,,	17 ,,	14 ,,	14 ,,	22 ,,	20 ,,
27 ,,	2 Aug.	6 June	24 ,,	21 ,,	21 ,,	29 ,,	27 ,,
3 Sept.	9 ,,	13 ,,	31 ,,	28 ,,	28 ,,	5 Nov.	4 Oct.
10 ,,	16 ,,	20 ,,	7 Feb.	4 Feb.	4 Jan.	12 ,,	11 ,,
17 ,,	23 ,,	27 ,,	14 ,,	11 ,,	11 ,,	19 ,,	18 ,,
24 ,,	30 ,,	4 July	21 ,,	18 ,,	18 ,,	26 ,,	25 ,,
1 Oct.	6 Sept.	11 ,,	28 ,,	25 ,,	25 ,,	3 Dec.	1 Nov.
8 ,,	13 ,,	18 ,,	7 Mar.	4 Mar.	1 Feb.	10 ,,	8 ,,
15 ,,	20 ,,	25 ,,	14 ,,	11 ,,	8 ,,	17 ,,	15 ,,
22 ,,	27 ,,	1 Aug.	21 ,,	18 ,,	15 ,,	24 ,,	22 ,,
29 ,,	4 Oct.	8 ,,	28 ,,	25 ,,	22 ,,	31 ,,	29 ,,
5 Nov.	11 ,,	15 ,,	4 April	1 April	1 Mar.	7 Jan.	6 Dec.
12 ,,	18 ,,	22 ,,	11 ,,	8 ,,	8 ,,	14 ,,	13 ,,
19 ,,	25 ,,	29 ,,	18 ,,	15 ,,	15 ,,	21 ,,	20 ,,
26 ,,	1 Nov.	5 Sept.	25 ,,	22 ,,	22 ,,	28 ,,	27 ,,
3 Dec.	8 ,,	12 ,,	2 May	29 ,,	29 ,,	4 Feb.	3 Jan.
10 ,,	15 ,,	19 ,,	9 ,,	6 May	5 April	11 ,,	10 ,,
17 ,,	22 ,,	26 ,,	16 ,,	13 ,,	12 ,,	18 ,,	17 ,,
24 ,,	29 ,,	3 Oct.	23 ,,	20 ,,	19 ,,	25 ,,	24 ,,
31 ,,	6 Dec.	10 ,,	30 ,,	27 ,,	26 ,,	4 Mar.	31 ,,

* Average length of gestation period, which, however, may be subject to slight variation in individual cases.

NSW Dept of Agriculture

When to join

The mating age for *heifers* depends upon both size and the breed. If the heifer is too small when joined, her body size may be permanently stunted and milk production during the resulting lactation will be reduced. She could also have more trouble calving

Early maturing breeds such as Jersey can be joined at a younger age than slower maturing breeds (e.g. Friesians). Jerseys can be joined at 15 months old while Friesians usually about 18-21 months (in terms of body weight about 230 kg for Jerseys and 300 kg for Friesians). Puberty occurs at approximately 2/3 mature body weight.

Cows should be mated about two months after calving to achieve a calving interval of twelve months. Any delay in rejoining may increase the time when the cow is dry and therefore unproductive.

Another point to consider is the time of the year you require the cow to calve. The gestation chart shows when to mate to achieve the desired calving date based on a gestation period of 283 days.

Conception

After puberty the normal non-pregnant cow ovulates every 21 days. The follicle on the ovary ruptures and the egg is transported down the oviduct. On the way down it may meet a sperm and implant in one horn of the uterus and be born as a nice healthy calf about 9½ months later. However things can go wrong.

Artificial Insemination (AI)

The advantages of A.I. are many — it eliminates the cost and worry of owning a bull, as well as the risk of venereal diseases, and it makes available the best quality semen from the best proven bulls from a wide selection of breeds. The cost will vary depending upon where you live and the chosen bull.

Most A.I. technicians carry a range of both beef and dairy bulls. A normal healthy bull produces thousands of millions of sperms at a single ejaculation. When collected and diluted there are sufficient sperms at each ejaculation to inseminate hundreds, and under special conditions, thousands of cows. Hence we need only keep the very best of bulls.

These are selected according to the merits of sire and dam, but as you cannot milk a bull, the only

way to know for certain that he is genetically superior is to see how his daughters turn out.

Artificial insemination — the technique itself

This is one of the most remarkable technological advances in modern agriculture. It has consisted of a number of important breakthroughs and discoveries such as:

1. The successful dilution of semen
2. The deep freezing of semen and its successful storage for many years
3. Devices for oestrus detection without the need for a bull
4. The perfection of the actual technique of inseminating

Regarding the latter, the following illustration may clarify.

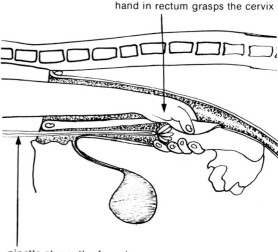

hand in rectum grasps the cervix

pipette at mouth of cervix

The technician is about to pass the pipette through the cervix. Once the pipette has been manipulated through the mouth of the cervix the semen is gently deposited just inside the body of the uterus.
Drawing by Graham Sharpe

In most dairying countries of the world the best young bulls each year are mated by A.I. to a random sample of females, and their daughters are compared when they come into milk in a few years time. Only those bulls whose daughters have the best production are kept and their semen is then released for widespread use. They are then known as 'proven' sires.

A typical sire survey for one artificial breeding centre

AUSTRALIAN BREEDING VALUES MAY 1989

Nasis Code	Bull Name	Sire	Maternal Grand Sire	Fat+ Prot	Milk Lts	Fat Kg	Fat %	Prot Kg	Prot %	Rel %	Type	D.C.	Cap	RU	F&L	MS	FU	RU	Size	$ per dose
GLENHAN	Glenalbas Enhancer Derek	Enhancer	Linmack	+73	+1088	+50	+.10	+23	-.21	56	NA									10.00
JOWAL	Jo-Wal Valiant Reward	Valiant	Astronaut	+64	+ 991	+37	-.07	+27	-.10	77	+8	+12	+5	+3	+6	+6	+2	+6	+5	15.00
TEEN	Otaheite Enhancer	Enhancer	Brundell Kriss King	+64	+1119	+40	-.10	+24	-.21	57	NA									10.00
LATT	Luccombe Enhancer Matt	Enhancer	Ned	+64	+1184	+42	-.11	+22	-.28	64	+8	+11	+7	+9	+5	+7	+4	+9	+4	15.00
JUPITER	Eurimbla Planet Jupiter	Starlite	Telstar Justin	+56	+ 798	+39	+.11	+17	-.16	82	-1	+3	-1	+1	-5	-1	-2	+2	+0	7.00
LINUS	Straight-Pine B.C. Linus	Chairman	Pete	+52	+1075	+28	-.27	+23	-.20	80	+7	+5	+4	+4	+8	+10	+11	+7	+6	12.00
MITESTA	Tunnybuc Mitestar	Superstar	Linmack	+44	+ 267	+34	+.45	+10	+.02	77	+4	+3	+0	-4	-2	+7	+6	+7	+2	8.00
OMER	Oberne Meadows Enh. Rocky	Enhancer	Skokie Prince	+44	+ 811	+29	-.07	+15	-.20	60	+4	+6	+2	+2	+4	+6	+5	+5	+1	7.00
TEMPLAR	Calleen Pansy Templar	Tempo	Linmack	+42	+ 536	+29	+.12	+13	-.09	70	+6	+4	+6	+6	+5	+4	+5	+3	+6	6.00
PACER	Parrabel Enhancer	Enhancer	Depositor	+38	+ 562	+23	+.00	+15	-.06	70	+6	+9	+8	+6	+1	+5	+3	+5	+8	6.00
MARSTON	Marston Rockman Progress	Tayside Pabs Rockman	Grifland	+37	+ 376	+25	+.18	+12	-.02	79	+6	+6	+0	+6	+4	+6	+5	+7	-2	7.50

A cursory glance at the figures above reveals that only bulls of above average production are listed, all other having been culled. It is interesting to note that the type characteristics (those relating to the animal's physical appearance) are also mostly positive. One would expect this if there is any relationship between form and function as discussed in Chapter One. The bulls are ranked according to their genetic ability to pass on the production of milk solids as measured by Fat plus Protein quantity with Glenalbas Enhancer Derek having the highest producing daughters! These would be expected to average around five hundred and forty-four litres of milk and 36½ kilo of fat plus protein above their average contemporaries milked under similar environmental conditions (same herd during same season). Complicated computer calculations known as Best Linear Unbiased Predictor (BLUP) are used to adjust measures of performance back to a common figure (Zero). What this means is that all the characteristics of the cow being mated to the bull and that of her relatives are considered in calculating the bull's ability so that if he is mated to a lot of above average cows he does not show up as better than he really is.

You will also note in looking at the survey of Glenhan that his reliability is only 56%. Perhaps this explains the lack of data at this stage on the physical characteristics of the daughters (overall type, body capacity, rump, feet and legs, mammary system, fore udder, rear udder and size). Utility characteristics such as temperament and milking speed may also be included in some sire surveys. The type figures quoted for Jowal indicate that his daughters will on average be the best of these bulls in overall general appearance (type) but in fore udder and mammary system the daughters of Linus should prove the best.

Remember that all these figures are measures of the bull's genetic merit. In laymen's terms it means that when he is mated to an average run of cows his daughters will average midway between his ability and that of their dams. In some earlier sire surveys the average was indexed at a hundred (as with Intelligence Quotient—IQ). Note that Marsten has a minus rating for size which means that he will produce daughters slightly smaller than the average. Finally it is always important to look at the reliability rating and figures for a bull like Glenhan with only 56% are much more likely to change as more information comes in about his daughters, than say for a bull like Jupiter with a reliability of 82%. The reliability figure is a measure of the size of the sample of daughters compared to the potential population of daughters. The bigger the sample compared to the total population the more accurate the production and type estimates are likely to be.

Problems with reproduction

Ovarian and hormonal malfunction, infection associated with difficult calvings and retained afterbirth, contagious diseasese and nutritional and mineral deficiencies are all capable of causing temporary or permanent infertility.

1. Ovarian and hormonal malfunction — cystic ovaries Sometimes the ovarian follicle fails to rupture and goes on producing oestrogen. This is known as a follicular cyst. The cow may exhibit almost continuous oestrus type behaviour (nymphomania) and does not go in calf.

A skilled vet may squeeze and rupture the follicle via the rectum, however, most vets usually prefer hormonal treatment using injections as there is always a risk of haemorrhage with mechanical rupture. The chances of a successful pregnancy following treatment are not 100%.

Other cows may ovulate or calve successfully but things go wrong when the *corpus luteum* or white body which develops on the ovary during anoestrus (between heats or during pregnancy) does not regress when it should. This is known as a luteal cyst. Again mechanical rupture or hormonal treatment may succeed and the animal commences cycling again.

A vet or A.I. technician should be consulted if your cow is not cycling regularly.

2. Environmental influences Poor nutrition may cause anoestrus or even abortion in early to mid pregnancy. Very hot conditions can also reduce fertility and conception rates and it can increase foetal deaths and abnormalities, especially if it occurs during the first weeks of pregnancy.

3. Diseases There are a number of venereal diseases of cattle. Some, like vaginitis, affect only conception. Others affect conception, implantation and/or later, foetal death.

Brucellosis, leptospirosis, vibriosis and trichomaniasis are the main ones found in Australia. Government veterinary officers and stock inspectors must be notified of any suspected out-break of these diseases. If a cow aborts, *particularly* between the third and seventh month of pregnancy, notify the Department of Agriculture immediately. Do not touch the aborted foetus.

At this point it may do well to remember that there are few cattle diseasese to worry about in Australia thanks to our quarantine laws and disease eradication schemes carried out by governments.

The dry period

Cows require a dry period of from 6 to 8 weeks to allow regeneration of milk secretory tissue in the udder and recovery of body condition lost during the lactation.

In a classic experiment the diagonal quarters of every cow in a herd were dried off 6-8 weeks prior to calving. The remaining two teats were milked right through to calving. These two teats only produced 60% of the milk in the subsequent lactation compared with the two quarters that had had an 8 week rest. As a tree sheds its leaves and grows new ones so a cow sheds her old milk secreting cells at the end of a lactation and grows another set during the dry period. If this process is

interfered with, production suffers in the following lactation.

Calving

Under Australian conditions it is not usually necessary to assist at the birth of any farm animals. Occasionally, however, one is called on either to give such assistance or to help a veterinary surgeon or someone else. It is therefore a good idea to know something of normal and abnormal births, and possible complications.

The female reproductive system consists of the uterus or womb, a sack-like organ which is connected through a round muscular opening called the cervix, to the exterior — by way of the vagina. This is a membranous canal also connected with the urinary system. The uterus is divided into two horns, which give rise to two narrow twisted tubes connected with the ovaries. These are suspended a little behind the kidneys in the abdominal cavity. The whole of this system is lined with a sensitive mucous membrane containing in parts numerous glands, which secrete fluid to lubricate the passage. The lining of the uterus is particularly suited for the nourishment of the unborn animal.

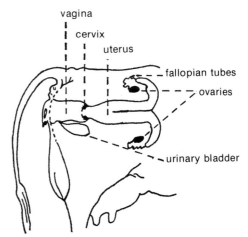

The genital organs of a cow.

NSW Dept of Agriculture

The approach of birth

The approach of birth is usually signified by several symptoms, varying in extent from one animal to another:

1. The udder, especially in young animals, usually becomes swollen and tender. The swelling might even extend along the abdomen and between the thighs. Some milk is usually present at this stage. The area around the tail head becomes loose and the vulva enlarges.

2. Usually there is a discharge of a thick opaque mucous (the cervical plug) from the vulva not long before the birth.

3. There is a notable change in behaviour. Cows wander off on their own and seek a secluded spot. They will stand up and lie down frequently and start to strain. The tail will start to hang out from the body and the "water bag" will appear and rupture. The front or hind legs of the calf will then usually appear. If nothing happens wthin a couple of hours from the onset of straining it may be well to investigate.

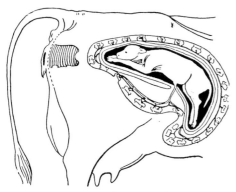

A normal presentation.
NSW Dept of Agriculture

The vast majority of calves are born normally from this position. Another position also considered normal is the 'posterior' one below.

Any positions other than these should be considered abnormal.

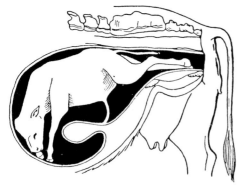

Hind-limbs and tail presenting.
NSW Dept of Agriculture

Investigating for abnormal presentation

Restrain the cow in a standing position by means of a halter or in a crush. Thoroughly wash hands and vulva of cow. Cut fingernails if long or use rubber or plastic gloves.

Investigation may be made via the rectum by feeling downward with the fingertips, or through the vulva. Any entry through the vulva should be as aseptic as possible.

If the front legs or head can be felt in the vagina in the normal position or the two rear legs in the normal posterior presentation, then assistance should be given immediately by securing a small chain, clean rope or binder twine around each presenting leg and then pulling backwards and downwards. Do not be frightened of pulling too hard or of hurting the calf or cow.

If two to three people cannot pull the calf or it is in some other position call a vet immediately you sense it is beyond your scope.

A normal calving sequence: The "water bag" appears.

The "water bag" has ruptured and the head and legs are ou

The calf is born (still largely within the foetal membranes). While many cows calve standing up, the majority would lie down.

The cow licks the calf clean and eats the membrane.

4 Milking and Milk Handling

The desirable characteristics of any routine husbandry operation are that it should be quick, effective and as pleasant as possible for both human and animal.

A good routine and a cow that is easy to get on with are the two essential ingredients for milking.

Our aim in milking is to get all the milk the cow has and get it as quickly and hygienically as possible, and in such a way as is pleasant for the cow and comfortable for the milker. In other words we aim to keep the cow, the milker and the drinker of the milk all satisfied.

Any chore carried out repeatedly should be designed to require as little regular problem-solving or awkward manipulation as possible. It should be something you have thought about and changed here and there so you can 'do it with your eyes shut'.

Some people are very good at such efficient organisation. Some dairy farmers by changing their routine or redesigning a yard or shed may have cut 10 seconds from the time it takes to prepare a cow for milking. If he milks 100 cows per day and saves 20 seconds/cow/day or 33 minutes/day or 200 hours/year, that is equal to five weeks labour at 40 hours/week.

There are, however, some essentials in a milking routine:

1. The udder should be clean.

2. The cow should be sufficiently stimulated to let down all her milk.

3. The utensils should be clean.

4. The milker should be comfortable and reasonably safe from injury.

5. The cow should be milked-out quickly.

6. There should be a minimal risk of mastitis infection during or immediately after milking.

7. There should be a minimum risk of dust and dirt falling into the bucket during the milking operation or rain running off the cow's back into the milk (with hairs and dirt).

These criteria can be achieved in a variety of ways.

Many people just milk the cow where she stands in the paddock. Others provide a little feed to encourage her not to wander off or get fidgety. At times you may just tie her up to a fence with a halter. If you leave a leather collar on your cow (with a fair sized metal ring on it) then all you need is a steel fence post in the ground and just drop the ring over the top of the post. Others build a shed with a head bail and feed trough.

If possible always get your cow standing up hill but on a bit of an angle so that you are right at the lowest corner. Why? Well, if your cow stands up hill her body (and udder) are thrown forward in comparison to her legs and it makes her back teats easier to get at.

It is also easier for the milker to be on the lower *side* of the cow as it is less awkward maintaining a balance without putting your legs under or behind the cow.

I keep two buckets for milking the cow (those strong plastic translucent ones with lids). One I milk into and the other I use for washing the cow and then sitting on. A lid is very handy if there is a likelihood of the bucket tipping over or milk sloshing out of it or being contaminated.

I have a plastic squirt bottle in the laundry with iodine dairy sanitiser in it.

My routine for paddock milking is:

1. Grab the milk bucket and add one squirt of iodine sanitiser.

2. Add warm water and swirl around milk bucket and transfer to washing bucket.

3. If I am hand feeding (autumn to early spring usually) I grab my feed container and head for the cow.

4. Wash and stimulate the cow. Cows late in lactation usually require a little more stimulation than others and Jerseys possibly more than Friesians. It is particularly important to get all the milk from the cow if you are only milking her once a day. Use a wettex or clean rag (hang in sun regularly) if cow is dirty, but a handful of water may be more hygienic on clean uddered cows (and it sanitises the hands). I usually wash from behind because I am standing up; you may prefer to sit on the side.

By now the udder, bucket and hands should all be sanitised and clean. It is a good idea to aim the first squirt of milk on the ground because it contains more bugs and will usually show up any clots if the cow has an udder infection (mastitis).

Some people milk wet i.e. with milk or petroleum jelly on the teats to lubricate them, and strip some of the time, others milk with the teats dry. The latter

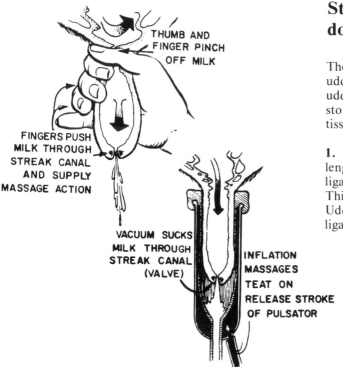

THUMB AND FINGER PINCH OFF MILK

FINGERS PUSH MILK THROUGH STREAK CANAL AND SUPPLY MASSAGE ACTION

VACUUM SUCKS MILK THROUGH STREAK CANAL (VALVE)

INFLATION MASSAGES TEAT ON RELEASE STROKE OF PULSATOR

Two ways to remove milk from the udder.
Babson Brothers Co., builders of SURGE

Structure of the udder, milk let-down and secretion

There are four separate quarters in the normal udder of a cow and these are not connected. The udder consists mainly of supportive, secretive and storage, circulatory, fat and often fibrous and scar tissue.

1. Supportive tissues The udder is divided lengthways into two halves by a strong central ligament called the *median suspensory ligament*. This ligament is the main support for the udder. Udders can become pendulous if the central ligaments, or front or rear attachments, are faulty.

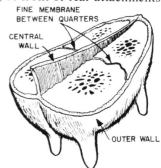

FINE MEMBRANE BETWEEN QUARTERS

CENTRAL WALL

OUTER WALL

The central suspensory membrane divides the udder into distinct halves. Very thin membranes divide the halves into quarters.
Babson Brothers Co.

is probably preferable though calves use the former method. Milking technique is difficult to learn from a book. The main thing to be achieved is to allow the teat to fill with milk and then apply pressure in such a way that the milk has to come out the bottom end and not slip back up into the udder again (and perhaps carry infection with it).

Milking technique towards the end of milking may also include working milk down from the udder into the teat. Cows vary in both their udder anatomy and physiology ('let-down' mechanism).

Some cows lend themselves to once a day milking quite satisfactorily, others dry off under that regime.

Many people run the calf with the cow say all day and lock the cow up at night and only milk in the morning or vice-versa and milk at night. The latter timing is preferable at least until the calf gets used to periodic separation and no longer incessantly bellows when taken from its mother. Neighbours may not take kindly to bellowing all night. Sometimes this scheme works, but sometimes the cow does not let-down too well and you may need to bring in the calf to suckle for a few seconds to bring on milk ejection. This is because she does not see the milker as a substitute calf when she already has a real one.

Milking routine is also mentioned on pages 29, 30, 32 and 63 (maintenance of health —mastitis).

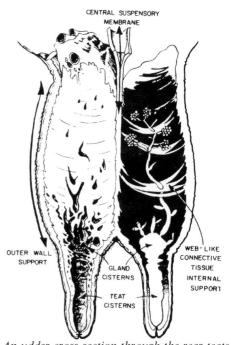

CENTRAL SUSPENSORY MEMBRANE

OUTER WALL SUPPORT

GLAND CISTERNS

TEAT CISTERNS

WEB-LIKE CONNECTIVE TISSUE INTERNAL SUPPORT

An udder cross section through the rear teats shows the suspensory tissues.
Babson Brothers Co.

. Secretive and storage tissue In most cows the two front glands produce about 40% of the milk and the two rear glands about 60%.

The opening of each teat is about ⅛" or 3 mm long (*streak canal*). It is held closed by a band of muscle (*sphincter muscle*). This sphincter muscle prevents dirt and foreign bodies entering the teat. Also, it must be able to withstand the pressure of milk as it builds up in the udder, or the teat will leak after a certain pressure is reached. However, this muscle must also be elastic if milk is to be withdrawn easily and quickly. Above the sphincter muscle the teat opens out. This is called the *teat canal* or *teat cistern*. Just above each teat is another cavity known as the *gland cistern*, which holds about 500 ml. From the gland cistern 8 to 12 main *milk ducts*, supported by connective tissue, lead upward, each one to a *separate lobe*. The main ducts branch off to smaller ones and this branching continues as the duct system is traced upward. The diameter of the ducts in the system is not uniform. They tend to enlarge just before joints for the purpose of milk storage.

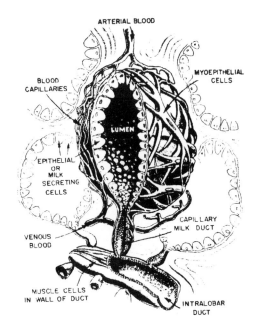

An alveolus showing secretory cells and allied tissues.
Babson Brothers Co.

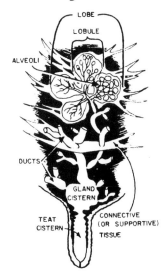

A complex duct system connects each teat to millions of alveoli. Alveoli make up lobules which in turn make up the lobes of each quarter. Babson Brothers Co.

Each of the glands is divided into *lobes* which are subdivided into a large number of *lobules* separated by connective tissue. Each lobule is made up of many hollow structures known as *alveoli*.

These alveoli are microscopic, balloon-like structures, each composed of a single layer of epithelial or skin cells. Each alveolus is in intimate contact with a rich blood, lymph and nerve supply and is covered with fibres of smooth muscle which contract and squeeze the milk out of the alveoli when 'let-down' occurs.

3. Circulatory tissue Blood is supplied from the heart to the udder via two large arteries, one coming down on each side and supplying half the udder. These pudic arteries in turn divide, thus making four rather large vessels, each of which supplies one quarter. By branching many times, the blood is dispersed through a very large number of tiny capillaries which supply the alveoli. The blood then flows on into the venous capillaries and veins.

Four large veins leave the mammary gland. Two of these are parallel with the two arteries and two extend forward along the underline (milk veins) and enter the abdominal cavity through openings (milk wells). Hence the venal (and circulatory) capacity of a cow cannot be judged with complete accuracy by referring to the size of her external 'milk' veins. These 'milk' veins are naturally varicose, i.e. have no functioning valves. Circulatory capacity and milk producing capacity are not unrelated. Approximately 400 litres of blood are required to pass through the udder in order to produce 1 litre of milk. Included with circulatory tissue is lymphatic tissue.

4. Fat is found around the edges of the gland where it is attached to the body and in any odd spaces to provide storage of energy and for protection.

5. Fibrous and scar tissue While some fibrous tissue is necessary supportive tissue, some cows develop 'meaty' udders. Occasionally Jersey and Ayrshire cows with prize winning show udders are found to be poor producers. Their large udders are fleshy and contain less secretive tissue.

Mastitis and injury can cause the development of lumpy and hard quarters, containing scar tissue which has replaced secretive tissue. Mastitis infection often does permanent damage to the udder.

Development of the udder and commencement of lactation

The udder is a skin gland and its secretive tissue actually grows from the outside inwards. Hence if extra teats are removed from a heifer when very young then no udder will develop to supply milk to the now non-existent teat.

Mammary glands commence development during the first three oestrus cycles when the heifer reaches puberty. There is, however, no further development until pregnancy. In the final phases of pregnancy lactation (milk secretion) occurs (see section on reproduction for more detail on the hormonal interrelationship which finally triggers parturition and lactation).

Milk secretion

This is quite a different process from milk ejection or let down. It is really a chemical process which goes on within the cells which make up the alveoli. It occurs as a result of a transfer of milk constituents and precursors (ingredients) from the blood capillaries outside the alveolus to the epithelial cells of the alveolar wall.

The more often a cow is milked per day, the higher her total production. This is partly due to the absence of pressure build up in the udder.

Milk let-down (milk expulsion or ejection)

Milk let-down occurs when the pituitary gland releases a hormone called oxytocin into the bloodstream. The hormone takes from 25 to 90 seconds (average 40) to reach the udder. Once there it causes the contraction of the muscle fibres surrounding each alveolus and the widening and shortening of ducts. When these muscle fibres contract the milk is forced out of the alveoli into the

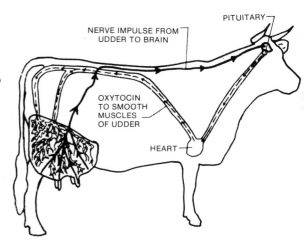

Diagram of the milk let-down reflex.

duct system of the udder and hence is readily available for withdrawal.

Oxytocin is a powerful hormone but has no lasting effect because it is destroyed by enzymes in from 2-10 minutes after entering the blood-stream. *Therefore, the cow must be milked immediately after let-down occurs or all the milk will not be obtainable.* If a cow receives a fright or suffers pain, a hormone is secreted into the blood-stream by the adrenal glands which are located near the kidneys. This particular hormone (adrenalin) causes the restriction of the blood supply to the skin and udder, it blocks the effect of oxytocin on alveoli, and increases the blood supply to the skeletal muscles. Hence the animal is prepared for flight or fight but not for milk let-down.

Cows are creatures of habit. If they suffer a stress or pain a number of times and under similar circumstances (reinforcement) they begin to associate the sights, sounds, smell, touch, etc. which they encounter at the time of stress with the stress itself. This is called a *conditioned reflex.* Such a cow may actually secrete adrenalin simply because similar circumstances have proved hurtful in the past. This is an example of how reproductive and survival drives can conflict.

Sight, and sounds, associated with pleasant or pleasurable experiences have the opposite effect. Hence some cows experience a partial let-down before they actually enter the milking bail. Similarly animals salivate when sight and sounds they have come to associate with feeding procedures are experienced.

This phenomenon of conditioned reflex is of great importance in the successful management of livestock. Good udder stimulation for 25-30 seconds helps to keep the conditioned reflex of milk ejection under the control of the farmer.

What stimulates the let-down hormone to be released?

The skin of the udder and particularly of the teat endings is sensitive to physical stimulation either by the mouth of the suckling animal or hand massage. When such stimulation takes place, the nervous system carries an impulse to the brain and the hormone oxytocin is released into the bloodstream.

Cows vary as to the amount of stimulation they require for positive and complete let-down. In many modern dairies the udder and teats are often subjected to little or no stimulation prior to the milking machine being attached. Cows which do not let-down show up as poor producers and are culled. Hence we have inadvertently selected against cows requiring very positive massaging of udder and teats to trigger let-down.

Only experience can show you if you are on the right track. Remember:

1. Handle cows quietly but firmly, and do not frighten or hurt them around milking time or in the vicinity of where you usually milk them. Cows definitely respond to quiet reassurance in the voice.

2. Cows late in lactation usually require more massage than fresh cows.

5 How a Milking Machine Works

Milking practice should follow an established pattern, which must be a pleasant one for the cow. It must be as convenient as possible for the operator, and of course, economical and hygienic, i.e. the machine design and operation is essentially one of optimising a combination of requirements which may tend to conflict. For example, a comfortable machine for the cow may prove inconvenient for the operator or difficult to clean; a design and operation which milks fast may cause greater mastitis. Wide bore-liners are known to milk fast and reduce cup slip and yet have, in some herds, greatly increased the mastitis incidence.

Hence, the continued research in the field of milking management.

How milk gets from the udder to the machine

Pressure of milk in the udder, and air pressure on the udder can cause milk to flow from the udder.

However, this does not normally occur unless the constricting force of the sphincter muscle at the end of the teat is overcome by either hand pressure or a vacuum being applied, to the teat.

Cows were first milked mechanically in England in 1819. The early designs were crude. The prototype of our modern machine was first used about 80 years ago. Since then there has been little basic change in the principles of machine milking. However, improved designs, innovations and more suitable materials used in the various components, particularly those introduced during the last 25 years, have greatly increased the efficiency and cleanliness of milking machines.

Milk is drawn from the teats of the cow by vacuum. When a strong vacuum is applied around the teat, the teat swells and the tension of the sphincter muscle is overcome, thereby opening the streak canal. Provided milk let-down has been achieved, the milk is drawn out of the teat cistern in a steady stream. Vacuum capable of raising a column of mercury 38 cm is regarded as the optimum for both speed and comfort. Some machines have a two vacuum system.

Why do we have a two chambered cup?

It is not hard to imagine that as the teat greatly increases in size when under vacuum the air pressure on the rest of the cow's body forces extra blood into the teat vascular system as well as milk into and out of the teat cistern.

When air rushes into the outer chamber of the teat cup, the liner or inflation collapses around the teat because there is air pressure on one side and vacuum on the other. This collapsing squeezes the teat and moves the blood past valves in the veins so that it cannot return and so, at the next expansion of the teat, fresh blood enters and cow comfort is maintained as per the following illustrations.

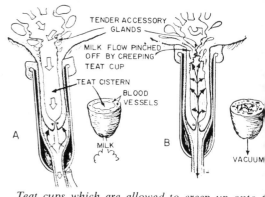

Teat cups which are allowed to creep up onto the udder can cause serious damage.

Babson Brothers C

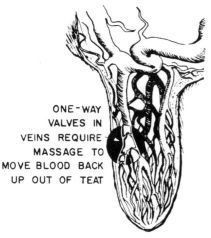

The circulatory system of an individual teat.

Babson Brothers Co

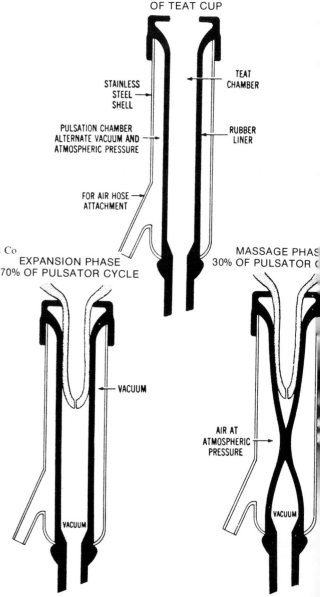

Diagrams of the teat cup at various phases of milking. During expansion phase (left) the rubber liner is held by vacuum in close association with the metal teat cup shell, and milk flow rate is maximal. During the massage phase (right) atmospheric air enters and collapses the ruber liner around the teat, and milk flow is minimal. There is constant vacuum in the teat chamber of the teat cup.

DeLaval Handbook of Milking, 1963, Alfa-Laval Inc.

What happens to the milk once it is in the machine?

The milk from each teat joins that from the other teats at the claw. Somewhere on the claw, or perhaps in each teat cup there will be a small hole called the *air admission hole*. This air rushes into the vacuum created by the vacuum pump and adds to that which has already sneaked in between liner and teat and at any other leakage points.

This air automatically rushes toward the vacuum pump where the air is constantly being removed, taking the milk with it.

Milking machine efficiency

The main factors to watch for are vacuum and pulsation stability.

1. Vacuum Stability

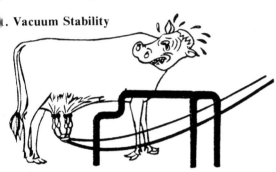

EXCESSIVE VACUUM

INADEQUATE VACUUM

Apart from the reasons illustrated vacuum instability usually means:

a) a slow down in milking as milk is not being removed as quickly as it might, quite apart from cups falling off.

b) an increase not only in discomfort but udder injury if vacuum levels go too high or cups are on for too long.

c) milk from one teat flowing back and reaching another — (impacts) — this increases the spread of mastitis. Milk may at times even flow from one cow to another cow in very faulty machines.

d) milk quality problems due to milk overflowing into parts of the machine where it isn't suppose to go, and also due to cups falling on the floor and sucking up dirt.

The vacuum level set (which is optimum for cow comfort and milking speed) will vary with the machine's design and may range from 44 kPA (12″ mercury) for a lowline to 50 kPA (15″ mercury) for a highline. This means the vacuum is sufficient to suck mercury up a tube 15 inches or 38 cms at atmospheric pressure.

Components of the machine which help achieve a stable vacuum are all part of a system of:

a) vacuum production (vacuum pump). There must be a sufficient volume of air being removed per minute.

b) vacuum storage (vacuum tank, releaser, milk line, vacuumised bucket).

c) sufficient capacity in the claw and milk line.

d) vacuum regulation (automatic regulator).

e) vacuum monitoring (a guage which checks on the whole system).

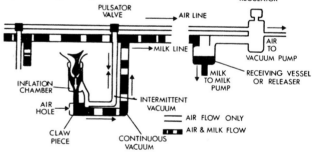

A schematic diagram of the lay-out of a pipe-line milking machine.

Of course, correct installation of the machine, correct slope in the milk line to prevent surges of milk, optimum pipe diameter and proper maintenance and adjustment of all these machine components is essential.

2. Stable pulsations at the teat

By stable is meant uniform from bail to bail and correct as to ratio, rate and degree of choke. All these settings are a sort of optimum or compromise to give the best of both worlds in regard to speed of milk removal and cow comfort.

a) *Pulsator ratio:* This is usually around 70% on milking phase and 30% on squeeze phase.

b) *Pulsator rate:* This is usually around 45-60 cycles per minute. A cycle is one squeeze phase and one release or milking phase.

c) *Choke:* Means the cushioning of the change from vacuum to air pressure in the outer chamber and vice versa. It is achieved by restricting or 'choking' the aperture as the milking and squeeze phases begin.

The following graphs taken by a vacuum recorder connected to the outer chamber illustrate rate, ratio and choke (from *Principles of Mechanical Milking* by W. G. Whittlestone).

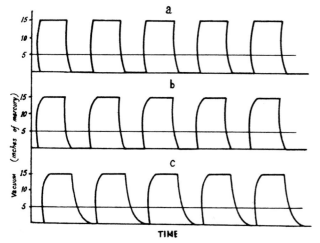

The effects of loading and choking a pulsator valve on the wave-form of the pulsator vacuum:

a) This is the unchoked vacuum graph measured at the pulsator valve.

b) This graph was obtained at the cups by inserting a take off nipple at the claw rubber.

c) A ⅛" choke was inserted into the pulsator air inlet port to obtain this graph.

The change in ratio measured along the 5" of mercury vacuum line is easily seen.

Other important factors to consider in efficient machine milking are:

a) some means of clearly identifying the end point of milking so that the cows are not overmilked.

b) sufficient liner or inflation length to ensure that the liner can close off below the teat during the squeeze phase. Remember a cow's teats may elongate to almost double their length when inside the cup of an operating milking machine (try the effect on your finger). Hence a 70 mm teat could stretch to almost 140 mm during milking. Therefore any liner which has an effective length of less than 140 mm is going to be too short for too many cows. Their teat orifices are not receiving proper respite from expansion that they should get during the squeeze phase. In the illustration of the squeeze phase on page 28 the liner has sufficient length.

This section on mechanical milking has been included because it has been found that some people prefer to milk by machine.

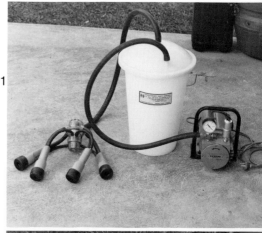

Milking machines suitable for the house cow owner. The p of these units depends on the type of claw, cups or bu used. 1. Oilless pump type. Very portable, will milk two c at once. 2. This unit milks up to four cows (or six goats once. 3. Oil type vacuum pump. Milks up to two cows at o Alfa-Laval Pty Ltd, Rydalmere NSW, supply a port milker for $1175 and Manus Nu-pulse, Dandenong Victc supply one for $1280 (both 1989 prices).

6 Milk Composition and Quality

The factors affecting milk quality and composition are of great importance to the commercial dairy farmer because of legal minimum standards and methods of payment based on quality and composition. Regulation of any such food industry is for the benefit of consumer and producer alike. As the house cow owner fits both categories, it is important to consider the valuable food constituents in milk, how these vary in the amounts present, and also what factors affect the keeping quality, flavour and disease carrying risk of milk.

Milk composition

The average analysis of cow's milk is as follows:

Constituent	Average (%)	Normal Range (%)
Water	87.25	82.6-89.5
Butterfat (B.F.)	3.75	2.0-7.0
Protein ⎫	3.2	2.8-4.4
Lactose ⎬(S.N.F.)	4.7	4.5-5.2
Minerals ⎭	0.75	0.6-0.8
Other constituents	0.25	

S.N.F. = Solids Not Fat

Water: Obviously the vehicle required for the constituents to remain in a suspended state. It is either in a free form, or bound by the protein or substances absorbed by the fat globules.

Butterfat (or milk fat): Forms the basic constituent of butter. It is lighter than milk and therefore rises to the surface when milk is allowed to stand. The fat is contained in globules which vary in size depending upon the breed; Jersey and Guernsey have the largest globules (better for butter making), Friesian and Ayrshire the smallest globules (better for cheese making). The larger fat globules rise to the surface more quickly when the milk is allowed to stand.

Protein: There are three proteins — casein, albumin and globulin. Casein is the most important due to its abundance (about 82% of milk protein), and also its capability of combining with calcium. Casein clots when the acidity is increased and when the enzyme rennin is added; this is important in the manufacture of cheese. Casein is also used in the manufacture of infant and invalid food, glues, fibres, paints, non-inflammable plastics and glossy paper manufacture. It contains virtually all the amino acids (protein building blocks) required by man.

Lactose: This is the principal carbohydrate in milk. It is a non-sweet sugar and is held in true solution. Lactose is converted into lactic acid by certain bacteria causing the milk to sour and finally curdle. This principle is used in ripening cream for butter making, milk for cheese production, and preparing yoghurts.

Ash or mineral matter: Includes all the important elements found in the body except iron, copper and perhaps manganese. They are either in true solution or associated with the milk solids. The main inorganic compounds are phosphates of calcium and chlorides of potassium and sodium.

Other constituents (minute fraction):

Vitamins: There are minute quantities of vitamin A, B, C, D, E and K in raw milk. Vitamin A is the most important and is either in the pure form, or as carotine which causes the yellow fat colour in Jersey and Guernsey milk. Vitamins A, D and E are fat soluble, B, C and K water soluble. Milk is a very important source of vitamins and minerals and is our principal source of calcium. If the cow is denied access to direct sunlight for long periods the vitamin D content in the milk will drop. Similarly milk vitamin A will drop if the cow is denied green hay or grass for long periods. Both these conditions are far more likely to occur in the northern hemisphere. The latter can occur in Australia when cows are fed solely on silage and concentrates during prolonged droughts. The vitamin C content of milk is usually insufficient for human requirements and is largely destroyed by pasteurisation.

However, milk remains as the outstanding and almost complete human food. Of all naturally occurring foods, we will survive on milk better and longer than anything else.

It is interesting to note that iron and copper, essential elements for animals, are missing from milk and one might wonder why. However, if we add them to milk they greatly speed up the oxidation of the fat which goes rancid and is unpalatable to man. Either evolution was very clever in knowing to keep those two minerals out of cow's milk so man could enjoy it, or God planned the cow especially for man's use — to be a dumb and peaceful friend, to learn from — should he condescend.

Milk also contains dead udder cells, white blood cells (leucocytes), bacteria and enzymes. These are all at a low level unless the cow is infected with mastitis (inflamation of the udder), or the milk is from a cow nearly dry or freshly calved.

Colostrum from freshly calved cows is significantly different from normal milk:

Water	71.5%	Serum proteins	16.0%
Butterfat	3.4%	Lactose	2.5%
Casein	4.8%	Ash	1.8%

Colostrum is high in globulin protein which is rich in antibodies conferring a degree of immunity to diseases on the calf. Casein is about the same level while fat and sugar levels are lower in colostrum. The milk is normal and fit for human consumption about seven days after calving.

Factors affecting milk composition

The lactose and ash content of milk is fairly constant over a wide range of breeds, management regimes and environments. However, the levels of butterfat and S.N.F. (due to variations in protein percentage) can alter appreciably due to a number of factors.

The factors which influence butterfat and solids content may be categorised as genetic and environmental.

1. Genetic
a) *Breed:* The following table illustrates the effect of breed.

Breed	Fat %	Protein %	S.N.F. %
Jersey	5.2	3.8	9.4
Guernsey	5.0	3.8	9.5
Ayrshire	4.0	3.4	8.9
Illawarra	3.9	3.4	8.8
Friesian	3.6	3.3	8.7

Generally speaking the higher the fat, the higher the S.N.F.; i.e. there is a positive correlation *genetically*. However, some of the *environmental* influences are negatively correlated so the interactions which occur are often somewhat complicated.

b) *Individuality:* Variation within breeds is as great as between breeds. Some cows vary greatly due to

changes in diet etc., while others are capable of remaining more constant. Certain family lines may be genetically higher in B.F. or S.N.F. This becomes an important factor in selecting replacement stock.

2. Environmental
a) *Nutrition:* This is by far the most important. The fat percentage is mainly influenced by the fibre content of the feed. If fibre in the diet is markedly reduced fat percentage may fall dramatically — even below 1% on rare occasions; however, there may be a corresponding rise in milk protein levels. Protein percentage (the principal constituent of the S.N.F. that varies) is mainly influenced by the amount of energy fed. See page 40.

However, feeding a cow lots of energy and digestable fibre does not necessarily ensure higher than normal butterfat and protein content in the milk as the acids produced in the rumen as a result of fermenting these feedstuffs also stimulate the production of lactose which increases milk volume production subsequently diluting the fat and protein down to more average levels.

Hence it is usually possible to *return* a cow to her genetical 'normal' level of milk fat and milk protein percentages by feeding an adequate ration, but abnormally high levels of butterfat and protein occur only when there is an unusual combination of circumstances which stimulate the production of fat and protein but not lactose. Fat, in milk, is in the form of an emulsion and protein is carried in suspension. These do not affect the strength of milk solution. Lactose, however, is a sugar and is dissolved in the milk in true solution and it is by far the most important constituent affecting the strength or osmotic pressure of the milk solution. This strength remains constant and in equilibrium with the blood. So, milk production follows lactose production and lactose production is related to energy consumption. So when cows capable of producing large volumes of milk receive insufficient energy, lactose production is affected less than protein and hence the S.N.F. content drops.

b) *Health:* Diseases such as mastitis can suppress milk yield and composition.

c) *Efficiency of milking:* The fore-milk is extremely low in B.F. (about 1%) and the strippings high (up to 10%). So complete milking will increase the B.F. percentage.

d) *Intervals between milking:* The longer the interval the lower the B.F. and S.N.F. *percentage* although the total *amount* of B.F. and S.N.F. may be higher.

e) *Yield of milk:* Generally high milk yielding cows have lower S.N.F. and B.F. percentage (dilution effect) but higher total production of S.N.F. and B.F. This interaction varies greatly from one animal to another.

f) *Season and weather:* Seasonal changes alter milk composition by their indirect effect on available feed and direct effect of ambient temperature. High temperatures increase B.F. percentage (by lowering total milk volume) but lower S.N.F. Cows eat less energy feed in hot weather. Low temperatures can also lead to raised butterfat levels.

g) *Age:* Both fat and S.N.F. percentage decreases after the first lactation at a rate of about 0.2% per lactation.

h) *Stage of lactation:* B.F. and S.N.F. percentage steadily decline over the first two to three months of lactation (not including the abnormal levels in colostrum). The lowest point corresponds with the cow's maximum milk yield point. B.F. then steadily rises until the cow is dried off. S.N.F. levels remain low until nearing the end of her lactation.

i) *Condition of the cow:* Cows calving in strong or fat condition produce milk with a higher concentration of solids than cows calving in poor condition.

j) *Excitement and oestrus:* Any factor which frightens or upsets the cow during or immediately prior to milking will prevent complete milking thus lowering the B.F. percentage. Cows on heat often spend reduced times grazing and ruminating and dissipate large amounts of nervous energy just fooling about.

As you can see there are a great many factors which can alter milk composition, and it is generally a combination of factors which cause a lowering of S.N.F. and/or B.F. The three most important factors, however, are breed, nutrition and stage of lactation.

Factors affecting milk quality

There is more to milk quality than just the chemical composition. Milk can become contaminated either whilst in the udder, during milking or while being stored, all of which can have a serious effect on quality. High quality milk should be free from dirt (sediment), taints or odours, pathogenic (disease) organisms, antibiotics and pesticides and should keep in a refrigerator for a number of days without souring (low population of spoiling organisms).

Contamination of milk in the udder

Milk may become tainted by the cow grazing certain types of feed. Silage, chou moellier (a cold climate fodder crop), rape, swedes, carrot weed, turnips and some vegetable scraps such as onion tops are the most common sources of feed taints. To prevent such taints, the cow should not be allowed to graze these types of feed for at least two hours prior to milking. Some cows, by nature of their peculiar digestion seem far more prone to produce tainted milk than others. If the milk is tainted allow it to stand in the bucket for half an hour before pouring and this will allow at least some of the taint to escape.

Diseases, particularly mastitis, can cause bacterial contamination of milk in the udder. Streptococci and staphylococci bacteria cause mastitis, and are in large numbers in mastitic milk causing a lowering of the keeping quality and producing a salty taste. Other bacteria such as tuberculosis bacilli and the organisms causing brucellosis (contagious abortion) may be found in milk from infected cows, thus creating a human health problem. Milk held in the teat itself usually has the highest contamination and so it is advisable to discard the first squirt. Cows drinking from dams or creeks where the teats can come in contact with contaminated water may cause bacterial contamination. Cows grazing recently sprayed plants or immediately after treatment for lice may produce milk containing unacceptable levels of pesticides.

The treatment of cows with antibiotics can also render their milk unfit for human consumption for a number of days. Antibiotics are a problem because:

a) Some people are hypersensitive to certain antibiotics and ingestion of even minute quantities can be serious or fatal.

b) Chronic exposure by humans to antibiotics lowers their effectiveness when later administered for a particular infection.

c) The presence of antibiotics in milk can prevent its utilisation for making cheese and yoghurt because the lactobacilli get knocked out by the antibiotics.

Most antibiotics for intra-mammary infusion are dyed blue so that by the time the dye is gone the milk is fit for human consumption. The milk in the untreated quarters is not affected providing the cow has not also received intra muscular (systemic) injections. In this case the milk, as with any 'blue' milk from mastitic cows should be used only for pets or calves.

Contamination during milking

For the house cow owner this would be the greatest source of contamination. Hand milking into an open bucket provides no protection against the entry of dirt, hair and bacteria into the milk. Clipped udders and well groomed cows do a lot to reduce hair and dirt falling into the bucket. Milking in dusty areas will cause an increase in the sediment level. The milk should be filtered immediately after milking to remove dirt and foreign material.

The milker *must* have clean hands as there is a continually high level of bacteria on hands and clothes which can contaminate milk.

Bacterial contamination will reduce the keeping quality, may cause a tainting of the milk such as burnt, bitter, putrid or fishy flavours, interfere with the milk's manufacturing properties or cause it to be unfit to drink owing to a high level of coliform bacteria from dung.

For those with a milking machine:

a) Always such through some rinse water (may be warm but not hot) immediately after milking operation is complete.

b) Suck through 5 litres of hot water containing the correct concentration of a proper dairy detergent at the recommended temperature (about 70° C).

c) Rinse through with boiling water.

d) Use an iodophor rinse prior to the next milking.

Alternative to the above routine is boiling water and sulphamic acid straight after milking followed by boiling water rinse.

One advantage of the first cleaning schedule mentioned is that the detergent can be used for other washing up and the iodophor rinse for disinfecting the cow's udder prior to milking, and the need to have boiling water is not as critical with the first mentioned cleaning regime.

The milking machine rubberware should also be checked regularly for cracks or perishing and the machine pulled down and inspected if some quality problem arises which seems to indicate a dirty machine.

It is because milking machines contain a number of joints and large surface areas exposed to milk and because they must be kept internally spotless that the single cow owner is better off mastering the art of hand milking.

Contamination during storage and processing

The main source of contamination during storage or processing is by bacteria on the surface of dirty containers or processing equipment such as a separator. Cleaning these utensils properly will eliminate this source. The utensils should be rinsed in cold water first to remove as much milk residue as possible; secondly, washed in hot water using a suitable detergent capable of dissolving milk proteins and emulsifying the fat; finally, rinsing in hot water to help sterilise and dry the equipment. Equipment may be chemically sterilised with a dairy sanitiser eg, "Idyne" or "Iosan" or a hypochlorite solution prior to its use, as an added precaution.

Rapid cooling and storage of milk at a temperature below 4° C is essential to prevent rapid multiplication of the bacteria already in the milk.

Stored milk can absorb taints from strong smelling substances such as mouldy food, disinfectants, fish, meat, onions, etc., so care should be taken during storage.

Milk can be contaminated from a wide variety of sources all of which affect its quality. However, by looking after the cow's health as well as paying particular attention to the hygiene of harvesting and storing the milk, a consistent high quality level can be maintained. In fact, with care and common sense raw milk from your cow, if cooled quickly, will keep as long, or longer than that which the milkman brings.

Disease risk from unpasteurised milk

Milk can carry a number of disease organisms such as those that cause diphtheria and sore throats. However, legislation regarding the pasteurisation of all milk was mainly to cover two diseases which have now been virtually eradicated from cattle in many of the more populous areas of Australia. These diseases are brucellosis and tuberculosis. Check the history and disease status of the cow when you purchase her and also the herd from which she comes. If she has a clean sheet the milk should not carry any disease risk.

There is little risk of food poisoning resulting from the ingestion of milk or milk products because the lactobacilli that naturally break down lactose to lactic acid tend to knock out most other bacteria. Hence the widespread use of milk in many parts of the world where hygiene is poor.

My three children were all raised on raw milk. However, during their first few months we heated any milk till almost boiling and cooled quickly. This treatment is more drastic than normal pasteurisation and kills all disease causing (pathogenic) bacteria.

7 Making Dairy Foods at Home

We have said earlier that milk is abundantly supplied with essential nutrients. Some people have claimed it to be the perfect food. The milk of any species may be ideally suited to nourishing its own young but it does not necessarily follow that it provides a perfect diet for adults of another species. Some people, mainly of non-European origin, who have not been accustomed to milk in their diets since infancy, cannot digest milk sugar — lactose; others are allergic to some protein fraction in cow's milk, or milk products such as cheese.

However, these problems apart, milk comes closer to being a complete food than any other non-formulated food which we know. There is no doubt that milk supplemented by a little fruit and fibre could sustain life and good health for a very long time.

Nevertheless, the same food, no matter how wholesome and nourishing, begins to pall when taken day after day without change. Everyone likes variety and milk can show its versatility in providing the basis for quite a range of attractive foods.

This has been recognised for thousands of years. Creams, butter, cheeses of many kinds and various fermented milks have been enjoyed by the human race virtually since creation.

The ancient arts of making these milk based foods have been reinforced by modern technology. The main benefits which this has brought about are:

a) Greater consistency in the quality of what has been produced — eliminating the periodic failures and hazards to health — spoilage and uncertain keeping life. In extending the keeping life the opportunity is presented of supplying a wider range to consumers and doing so over a longer period.

b) Greater economy and efficiency of production, elimination of waste — production at minimum cost, offering the opportunity to maintain low prices for the consumer and reasonable returns for the producer.

c) In some cases it has extended the range and variety of foods available.

To achieve all this the people working in the dairy manufacturing industry have undergone a long and exacting training. They are professionals and employ hard won skills and sophisticated techniques to achieve products of high and consistent quality and to offer them at the most competitive prices.

It would be folly to imagine that raw amateurs can automatically match their performance. Superficially it might appear that a home made product should cost less but this usually discounts all the costs which go into the operation — especially the value of the labour of the home operator. There is no magic in home made foods; the best, consistent quality and value for money is usually obtained through competitive, commercial channels. The main advantages offered by 'home made' foods, however, are:

a) Utilising excess milk. Most of the time the average cow produces more milk than a family can normally consume in the liquid form.

b) They can be given any special characteristics or particular distinguishing touches desired by the family.

c) As they are usually consumed very shortly after they are made there should be less chance of deterioration; protective packaging and distribution safeguards, which can be expensive, become of minor importance.

d) Making things at home can be fun. However, as in all things, success is the reward of effort, understanding and meticulous attention to detail.

Dairy foods derived from milk often keep better than liquid milk itself, hence they can be made when the milk is abundant, and stored for later in the lactation when the cow may only just be meeting the liquid milk requirements of those who 'operate' her.

The processing and keeping of food

Down through the ages spoilage of food has been delayed by:

    a) removing water (dehydration)
    b) adding salt
    c) adding sugar
    d) heat sterilisation
    e) raising acidity (pickling)
    f) keeping it cool.

If you think of any processed food or dairy product, one or more of the above treatments has been applied unless the food, such as a dry grain contains little or no water or oil. Being in a liquid form and being a balanced food, milk is an ideal breeding ground and food for bacteria. To reduce the level of bacterial activity in milk we can:

a) dry it (butter, cheese, milk powder)
b) add salt (cheese and butter)
c) add sugar (condensed milk, ice cream, etc.)
d) heat sterilise (U.H.T. milk, evaporated milk)
e) raise acid levels (cheese, yoghurt, etc.)
f) refrigerate. This applies to most dairy products and greatly reduces the rate of bacterial activity.

Complete vacuum sealing following sterilisation is also used in modern food technology but was not available to man until modern times and not usually applied in the home preservation of dairy foods though it is widely used in bottling fruit.

Handling whole milk

Immediately after milking, pour the milk through a plastic kitchen strainer about 140 mm diameter into which has been placed a proper milk strainer pad or piece of clean cheesecloth. A cheaper but more fiddly operation is to use half a paper tissue. Once milk has cooled and the fat has risen it will not strain very effectively and some cream will be lost. We have found 5 litre billies or tall 2.5 litre plastic jugs best. These take up less room in the refrigerator and are very easy to skim.

If you milk a Friesian, Illawarra or Ayrshire you may prefer to drink the milk straight. However, many people with high butterfat cows prefer to skim off most of the cream. Remember, rule number one in all dairy processing operations: constant and thorough attention to detail in all matters of hygiene.

Skimming cream

As a rule the house cow owner need not bother with owning a cream separator, especially if he milks a breed (Jersey or Guernsey) which produces large fat globules that rise relatively quickly.

The poured and filtered warm milk should be refrigerated immediately and allowed to stand for 24-48 hours. Use a stainless steel or plastic table spoon (no brass, iron or copper) to skim off the cream.

This operation can produce quite thick cream of excellent quality. It is best stored in glass jars with plastic lids. Well washed peanut butter and jam jars are excellent but avoid pickle jars as they can carry a lingering smell which could taint the cream.

Hand turned cream separators are not cheap (around $300 for the cheapest new one in 1982) and are a real fiddle to wash up and keep clean. Early models not made of stainless steel are a real risk as

the tinning has usually worn off the discs, thus exposing the milk to iron surfaces.

Making butter

This is an inverse phase change. Cream consists of fat globules suspended in a water solution as an emulsion. Butter consists of water droplets suspended in a fat solution.

To bring about this change the cream is agitated causing the lipo-protein 'skin' of the fat globules to break down as they are 'bashed' together. Once this droplet membrane is gone the fat particles coalesce into larger globs of fat.

Ordinary refrigerator temperature is O.K. for making butter. Do not use warm cream.

Small butter churns are available (usually second hand) but a kitchen cake mixer is quite acceptable providing you do not turn the speed up too high and are not too ambitious as to the amount you wish to churn at once otherwise you may end up with butter splattered around the kitchen. If you see yourself making quite a bit of butter a small hand churn may be a good investment as you can make more butter at a single operation with less risk of making a mess.

Energetic low technology types can try a large jar half full of cream and shaken vigorously or tied to some mechanical contraption.

The cream is beaten, until first it whips, then 'breaks'. The buttermilk is drained and can be used for baking or other purposes, and the butter carefully worked with pats to remove more water and to achieve a smooth spreadable texture. Most people find a little salt added improves the flavour and this also helps to preserve the butter: about 20 grams is needed for each kilogram of butter.

Butter contains about 15% moisture and 1½% solids not fat. Allowing for 2% salt the finished butter will contain some 81½% fat.

Making cheese

Cheese is made by treating milk:
1. with a special bacterial culture (which turns part of the lactose to lactic acid, giving a pleasantly sharp tangy flavour), and
2. with rennet which causes the milk to set like a junket. When sufficiently firm the set milk is cut into small cubes and separates into curds and whey.

Different varieties of cheese depend on:
a) the amount of moisture retained in the curd.
b) the acidity developed,
c) the physical nature of the curd, as determined by

the manner in which it is manipulated during the cheese making process,
d) the condition and length of the time for storage after the curd has been drained, pressed and wrapped.

Cheddar cheese

Pasteurise milk then cool to 30°C. Stir in starter culture, ¼ litre for 10 litres of milk. Stand 15 minutes or to desired acidity. Stir in a little rennet, ½ teaspoon to 1/3 cup water to 10 litres milk. Stand undisturbed until milk 'sets' — 30-50 minutes. Cut curd (junket) into 2 cm cubes. Stand 10 minutes. Start to very slowly raise temperature (1°C in 3 minutes until 30°C) and curd shrinks and becomes firm. Ladle curd into cheese cloths, tie and hang up to drain for 30 minutes. Cut into chips and add 30 gms salt to 1 kg curd (about 3%) (1 kg curd from 10 litres milk). Line mould with cheese cloth and fill with curd and pack tightly, fold cloth over top, place top on mould and press. Unwrap next day and ripen at room temperature (20°C) for two days (to harden rind) then dip in wax or re-wrap tightly in plastic and store at 10-12°C until desired stage of maturity.

Cottage cheese

Pasteurise skim milk at 70°C. Cool to 30°C. Add starter 50 ml/litre. Stand until sets (overnight). Cut into cubes *carefully*. *Slowly* heat to 49°C, careful not to shatter curd. Drain, wash three times with water. Warm water first, then cooler, then cool. If water not ideal add a few drops of 'Milton' or some other similar sanitiser. Drain in cheese cloth bag overnight. Add cream and salt.

The Australian Dairy Corporation, which has offices in most capital cities, has a large range of dairy food recipes available.

Leaflets describing the manufacture of some dairy foods in the home are also available from some State Departments of Agriculture.

Making ice cream

See following chart.

Home-made ice cream recipes

	ice milk		ice cream					
recipe no.	1	2	3	4	5	6	7	8
whole milk	1 Litre	¾ Litre	600 ml	300 ml	600 ml	300 ml		600 ml
cream (30% butterfat)	—	—	300 ml	300 ml	300 ml	300 ml	600 ml	300 ml
skim milk powder	1/3 cup (55 g.)	—	1/3 cup (60 g.)	—	½ cup (85 g.)	—	—	—
evaporated milk unsweetened (410 gram can)	—	1 can	—	1 can	—	2 cans	2 cans	—
sugar	¾ cup (180 g)	200 g	¾ cup (170 g)	¾ cup (180 g)	¾ cup (170 g)	¾ cup (180 g)	1 cup (250 g)	¾ cup (180g)
gelatine	◁———————— 2 tsp. (6 grams.) ————————▷							
eggs	—	—	—	—	—	—	—	four
flavouring	◁———————— see below ————————▷							

Recipes 5 and 6 contain extra milk solids and are more suitable for ice cream whipped up in a cake mixer and statically frozen. This ice cream should not be stored for long.

Recipe 7 is for an extra rich ice cream. Variations are possible by changing the proportion of milk and cream in receipes 6 and 7.

Flavouring: (1) Vanilla ¼ to ½ tsp. vanilla extract to each batch. (2) Fruit, prepare a fruit puree by lightly cooking 2 parts fruit to 1 part sugar and adding 10-20% of the puree to the basic ice cream mix. *Chocolate:* make a syrup using 10-20 grams fine cocoa plus an equal weight of sugar dispersed in 1 cup boiling water; add this to the basic mix before freezing. *Coffee:* make a coffee syrup in a similar way using 5-10 grams. instant coffee.

Procedure: mix the milk, cream and evaporated milk (where used) in a saucepan placed inside a larger pan containing water, on a stove. Heat with gentle stirring until the temperature reaches 40°C. Mix the sugar and fine granular gelatine (and skim milk powder if used) and add slowly to the warm liquid mix stirring vigorously until it is all dispersed. Continue heating until the temperature reaches 75°C, hold at this temperature for 10 minutes then cool rapidly by standing the pan containing the mix in a sink of cold water, finally cooling by adding ice to the water. Stand the pan of mix in a refrigerator overnight before freezing.

Freezing: is best done in an ice cream freezer which is a rotatable metal cylinder with a scraper set in a tub of ice and salt (1 cup of course salt to each 10 lb. crushed ice). If no freezer is available the mixture may be whipped with a cake mixer and frozen in the deep freeze compartment of the refrigerator. This is a less satisfactory procedure and is very tedious as the whipping needs to be repeated several times whilst the mixture is freezing.

Making yoghurt

Heat milk to 85-90° C and hold at this for about 15 minutes to kill all bacteria. Cool quickly and hold at 42-44° C in a Crock-Pot, electric fry pan, large thermos flask, yoghurt maker, etc. Add natural yoghurt or culture as soon as temperature drops to about 42° C. Hold till desired thickness (about 6-10 hours). Put in jars 2/3 full and put in fridge. Shake when cold to cream up yoghurt and fill remaining third with desired fruit, honey, sesame seeds, etc. if required.

Variations on this recipe include:

a) Testing the temperature with a clean finger. If you can keep it in the milk for 10 seconds it is cool enough.

b) Adding some skim milk powder when you add the natural yoghurt.

c) Using a proper starter culture in place of natural yoghurt. This can be stored in frozen pellet form in liquid nitrogen or obtained fresh from a cheese or yoghurt factory, or purchased in powder or tablet form from health food shops.

d) Keeping the bowl warm using a quilt, blanket or towel.

Some yoghurt recipes

Yoghurt cheesecake

We have often made this and it is truly hard to beat. Take one 18 cm (7") flan case made from two cups of crushed biscuits (sweet), 1 teaspoon mixed spice, 112 g (4 ozs) melted butter. Combine and press into a well-buttered pie flan.

Filling: 1 carton (220 g) plain yoghurt, 3 tablespoons white sugar, 1 tablespoon currants, 1 teaspoon grated lemon rind, 1 egg, grated nutmeg.

Beat egg and sugar together and then add yoghurt, currants and lemon rind. Mix well, spoon into flan and sprinkle with nutmeg. Cook in moderate oven for ½ hour or until set.

Raita alu (potato salad)

Chop boiled potato into cubes and add chopped onion, garlic, mint and natural yoghurt to taste and stir in. Some people also like yoghurt with cucumber and cumin.

Yoghurt custard

340 gms (12 ozs) plain yoghurt
110 gms (4 ozs) water
2 eggs
1 tablespoon sugar
½ teaspoon vanilla essence
nutmeg

Pre-heat oven to low to moderate heat. Whisk together yoghurt, water, eggs, sugar and vanilla essence. Pour into a buttered pie dish, sprinkle with nutmeg. Place in a dish of water and bake for 1 hour. Yields 4 serves.

Lemon crunch pie with plain yoghurt

225 gms (½ lb) Honey Snap biscuits
110 gms (4 ozs) butter
225 gms (½ lb) plain yoghurt
½ can sweetened condensed milk
4 tablespoons lemon juice

Crush biscuits with a rolling pin. Melt butter and mix with the biscuits, blending well. Press firmly into a buttered 18 cm (7") tart plate to form a shell. Pour condensed milk into basin and fold in yoghurt, lemon juice and rind. When beginning to thicken, pour into shell. Chill in refrigerator for 12 hours. Before serving, decorate with whipped cream.

Yoghurt salad dressing

Rub salad bowl with garlic. Place two parts of olive oil or salad oil, 1 part of lemon juice (or white vinegar) salt and pepper to taste into salad bowl. Cut up medium sized onion finely. Let stand (alternatively add grated "Icebert" cheese), then add yoghurt equal to volume of other liquids.

Yoghurt salad dressing for slimmers

Mix together desired quantity plain yoghurt, ¼ teaspoon sugar, ⅛ teaspoon of salt, ½ teaspoon lemon juice and the grated rind of a quarter lemon.

Veal paprika with plain yoghurt

1 kg boneless stewing veal	1 level teaspoon salt
2 medium onions	1 cup tomato sauce
good squeeze lemon juice	1 teaspoon chopped parsley
1 level teaspoon mustard	2 tablespoons plain flour
1 clove of garlic	3 tablespoons melted butter
½ cup mushrooms	1 tablespoon paprika
½ cup plain yoghurt	

Cut veal into 1" cubes and roll in a mixture of flour, salt, pepper and mustard. Heat the butter in a pan and fry onion and garlic until brown. Lift out. Fry the veal until lightly brown. Transfer the meat and onion to casserole. Add a little more butter to pan and stir in another tablespoon of flour. Cook until brown. Add a cup of water or stock, the tomato puree and lemon juice and stir until the liquid boils and thickens. Pour over veal in casserole. Add paprika and butter mushrooms, cover and cook in a moderate oven for 1-1½ hours or until veal is tender. Just before serving, spoon the yoghurt on top and sprinkle with chopped parsley.

Yoghurt pancakes

225 gms (8 ozs) self raising flour   pinch of salt
2 tablespoons sugar   1 egg
300 ml (½ pint) of water
150 ml (¼ pint) of plain yoghurt

Sift flour, salt and sugar into a bowl. Combine egg, water and yoghurt and beat into sifted dry ingredients. Spoon into (3") 80 mm rounds on buttered griddle and cook until golden brown on both sides. Layer three together on plate joined with raspberry jam and yoghurt. Yields 4 serves.

As can be seen there are innumerable recipe variations based on yoghurt. One popular one in our house is plain yoghurt whipped into an ordinary jelly.

Basic cream cake mix (saves making butter first)

1 cup cream, 2 eggs, 1½ cups plain flour, 1 cup castor sugar, 2 teaspoons baking powder, pinch salt, vanilla essence.

Whip cream in basin until soft peaks form. Add eggs one at a time, beating well after each addition. Sift flour, baking powder, salt, and sugar. Add dry ingredients to cream mixture, stir until well mixed. Pour into greased and floured 8 inch spring-form pan. Bake in moderate oven for 45 minutes.

Vanilla meringue ice cream

2 egg whites, ¼ cup sugar, ½ teaspoon vanilla, 1 cup heavy cream (whip), ⅛ teaspoon salt.

Beat egg whites with salt until frothy, then beat in the sugar, gradually, to make a stiff meringue. Whip cream until thick enough to hold a soft peak, add vanilla and fold into meringue. Pour into freezing tray and freeze, without stirring until firm. (Additional ingredients — 2 bananas, 4 passionfruits, desertspoon lemon — mash into bananas and add.)

Lemon Cream Sherbet

4 cups milk   2½ cups sugar
2 cups light cream   2 cups lemon juice

Heat two cups milk to scalding. Remove from heat. Add sugar and stir until dissolved. Add remaining ingredients. Freeze in ice cream freezer or in freezer section of refrigerator. If frozen in ice cube trays, the mixture should be partially frozen, then removed, beaten until fluffy and returned to complete the freezing.

The addition of a dessertspoon of gelatine (dissolved in a little hot water) before the second beating will improve the stability and consistency of the mixture.

8   Feeding the House Cow

An important key to success in keeping a house cow is in knowing what to feed her in order to keep her healthy, productive (milk) and fertile.

The cheapest feed producing these desired results is of course the best.

A combination of theory and experience enables a livestock manager to utilise his labour and feed resources in the best possible manner.

How a ruminant utilises and digests food

A cow has been called a 'walking fermentation vat'. It is like the old crude oil tractor — it runs on cheap fuel but loses a lot in the exhaust (belching and faeces).

It can be seen from the diagram above that not a

What happens to food energy

food eaten = gross energy (starch cellulose, fats and some protein)

30-40% faeces ◄——

digested energy (T.D.N. or Total Digestible Nutrients)

6% methane (belched) ◄——

absorbed energy

6% excreted as urine ◄——

metabolisable (useful) energy (ME)

25-35% *heat production* from rumen, and nutrient metabolism (wasted and even a burden in hot weather but normally useful for maintaining body temperature)

net energy — 10-30% for products, work, cell maintenance, and growth (except during extremely cold weather when all feed energy may be channelled into survival) great deal of ruminant gross energy consumption can be available for productive purposes, such as milk, growth and maintenance. As a cow dries off, the net energy available may drop below the 20% mark as feed conversion to body tissue is less efficient than feed conversion to milk and dry cows are often on poorer quality feed. This appears to be somewhat inefficient but one must recall that as a living system ruminants have survived and are among the earth's most numerous mammals.

Their heat loss also appears to indicate some inefficiency but this is inevitable in all warm blooded animals. Again warm blooded animals, though requiring many times the maintenance ration than for similar sized cold blooded animals, have distinct advantages in that they can govern their body metabolism independently of environmental temperature.

A cow's appetite increases in very cold weather and decreases in hot weather, particularly if it is also very humid. Therefore, in hot weather a cow tends to curtail feed consumption and milk production in order to reduce heat production due to rumen fermentation and the metabolism of nutrients. Concentrates produce less 'waste' heat than do roughages. In cold weather rugging can save feed by reducing heat loss.

How the rumen works

The cow usually eats grass or roughage by quickly chewing it lightly and mixing it with copious quantities of saliva as she swallows. Some hour or hours later she will stand or sit quietly and commence regurgitating her cud and chewing each ball 40 to 60 times and then reswallowing. The rumen contracts about every half minute and this moves fresh coarse fibre to the mouth of the oesophagus for regurgitating and causes the finer particles to settle further towards the bottom of the rumen contents. It is the rumen liquor and finely broken up particles that are washed into the third stomach (omasum) where a lot of the water is

removed and then on into the fourth stomach and the rest of the alimentary tract.

All the time the feed is in the rumen cellulytic or fibre attacking bacteria are breaking down the roughage and releasing energy in the form of methane gas, heat, and fatty acids which are absorbed by the blood through the rumen wall. Some of these acids are converted by the body to glucose, some to fat and some are used as they are.

Nitrogen from soluble protein and non-protein nitrogen (e.g. urea) is usually utilised by bacteria in the rumen to form high quality microbial protein and a small amount can be absorbed through the rumen wall as urea.

Insoluble protein and microbial protein are broken down in the fourth stomach (abomasum) into amino acids ready for digestion.

It is very important to have some understanding of rumen function as it may save you problems.

Cows were never designed to run on refined fuel and require some roughage (at least around 30%) in order to avoid rumen malfunction. In other words, the rumen does not work properly without it. The fibre should not be chopped too fine as it will drift down in the rumen liquor and pass through before the bacteria have had time to act on it thoroughly.

Another important aspect of ruminant physiology is the maintenance of *stability* in the ruminant environment and ecosystem.

Violent changes in feed type can upset a cow because different bacteria break down different raw materials. Other bacteria act on the products of an earlier stage of fermentation. Still other organisms prey on the beneficial bacteria and keep their numbers in check so that they do not multiply too rapidly and produce more lactic acid, for example, than other bacteria can cope with at any one time. If such an event should occur as when a cow eats a lot of starch to which she is not accustomed (e.g. grain), then more acid will be produced than can be coped with, the acidity of the rumen will rise, changing the internal environment and upsetting the whole balance of organisms as some of them tolerate acid conditions better than others.

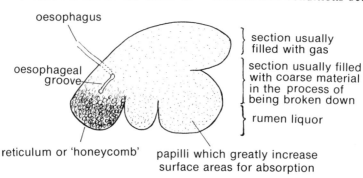

oesophagus

oesophageal groove

} section usually filled with gas

} section usually filled with coarse material in the process of being broken down

} rumen liquor

reticulum or 'honeycomb'

papilli which greatly increase surface areas for absorption

Section through a cow's rumen from the left side.

Other reasons why cows can get indigestion from eating too much grain are:

1. They mix a lot less saliva with it as it is not regurgitated and tends to fall quickly toward the rumen floor. Saliva contains urea and bicarbonate and is alkaline. Cows produce 60 to 160 litres of saliva per day (an astonishing amount).

2. Once a cow gets indigestion (usually caused by high acidity as in humans) her rumen may slow down or cease to contract thus inhibiting regurgitation and salivation.

If you discover that your cow "has got into the chook food" then drench her with 500 g of bicarb of soda as a precaution. This will help reduce rumen acidity. If she is already really sick, call your vet, as she could die if she does not receive expert help. (See section on ketosis on page 62.)

Cows were designed for grass but can successfully cope with a very wide range of feedstuffs. A good rule of thumb is not to give too much of anything and introduce changes gradually (over a period of a week or so).

Bloat is another rumen malfunction. While it can occur with a variety of feedstuffs it is mostly associated with hungry cows rapidly ingesting pasture dominant in succulent legumes such as clovers and lucerne. The barrel of the cow becomes greatly distended even to above the backbone on the left side. Bloat may be fatal. Cows often eat rapidly when put onto fresh pasture. The legumes produce a lot of gas as fermentation commences but the tiny bubbles do not coalesce into larger bubbles rapidly due to the particularly high surface tension of the froth. However, if the cow is not particularly hungry, is well conditioned to bloaty pastures (and some cows learn better how to cope than others) or is taken back out of the pasture before she eats too much, then the cow may just swell up slightly on her left side. If she stands with her front feet up hill at this stage she will tilt the gas pocket forward and start belching.

If the cow eats beyond a certain amount, the internal pressure becomes so great that she is not able to open the lower end of her oesophagus as a reflex action and allow gas to escape, and as the rumen is often almost completely full with feed and tiny bubbles, there is very little straight gas for her to belch out anyway.

Quick action is needed if the cow is up very high on the left side and grunting. If she is not too bad drench her with a proprietary anti-bloat medication, or 600 ml of cooking oil or linseed oil and stand her up hill (see p. 15 section on drenching). An alternative is to inject one of the above substances directly into the rumen (same spot as for stabbing).

If she is struggling to breathe and starting to

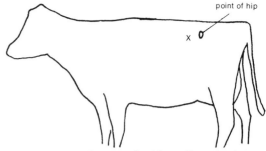

Where to stab a cow for bloat. X marks the spot.

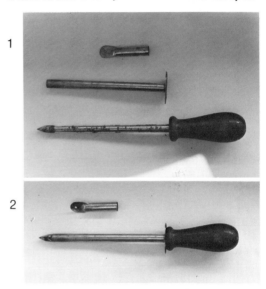

A trocar and canula. 1. Separate 2. Together and ready for use (with cap removed).

stagger, stab her at once with a trocar and canula, or anything sharp as not a moment's time can be lost. Usually once a cow goes down she is dead and what a ghastly death.

If the cow is stabbed then it is a good idea to inject her intramuscularly with antibiotics to help avoid the possibility of peritonitis which is often fatal.

Food requirements of cattle

The clever manager is the one who can best match the feeds available to the animal needs and who thinks ahead in this regard. Apart from containing the right amounts and balance of nutrients and containing some bulk or roughage as previously referred to, the food must also be palatable and non-toxic. Within these guidelines cattle can *exist* on an extremely wide range of diets — from molasses — shredded paper — urea combination through to whole live fish. I emphasise, they can *exist*, but do not necessarily thrive on extreme diets.

In order to meet the cow's needs for milk production, regular calving and good health, her energy, protein and mineral intake must be *sufficient*, and *balanced*.

Some people have divided cows needs into:

1. Those sufficient to *maintain* an animal in health at a static bodyweight, and
2. Those needed in addition for any growth, late pregnancy (foetus) or *production* needs.

It is obvious that maintenance needs are not absolutely static but vary with environmental temperature and any accompanying wind and/or rain, distance walked to grazing, and of course body weight.

The requirements for maintenance for a 1000 lb or 454 kg cow have been traditionally stated as 2.7 kg or 6 lb of total digestible nutrients of which 10% is protein. In terms of metabolisable energy (M.E.) the same sized cow would require 49 megajoules (MJ) or 12.3 megacalories (Mcal). In Australia megajoules have been adopted as the standard measure. Feed *energy* value is then measured as so many megajoules per kg drymatter — M/D being the shorthand symbol.

Apart from minerals and vitamins which make up only a minute but important fraction, food consists of energy in the form of fats, starch, sugars, cellulose and protein. However, the body only uses protein *for energy* if it is in excess of needs for maintenance of body tissue, growth, milk production, etc., or if energy is lacking elsewhere in the diet. In other words protein can be used for energy but carbohydrates and fats cannot be turned into protein by mammals though some bacteria including some living in the rumen, can do it under certain circumstances and thus make microbial protein for the cow from raw materials in her feed.

Approximately 3.2 kg of Digestible Nutrients (T.D.N.) of which 20% is protein are needed by the cow in addition to maintenance needs for each 10 litres of 4.0% Fat Corrected Milk (FCM) produced.

A *maintenance ration* is the 'rent' you pay the cow for the use of her body — a 'retainer' if you like.

A *production ration* is the 'wages' you pay the cow for work — 'commission' so to speak.

Early in lactation good dairy cows 'borrow on the rent' by losing body weight and converting these body reserves into extra milk. Later in the lactation and during the dry period a good manager will try and allow her enough feed to gain weight and 'pay back the rent'.

As it will be seen from the graphs, if we feed a cow well late in lactation, i.e. give her a sudden boost, it will result in only a marginal improvement in production with the bulk of the feed being converted to body tissue (the cow gains weight). Conversely, early in lactation a good cow responds to improved feeding mainly by increasing her milk yield with only a marginal effect on body weight; she is also likely to come on heat sooner and go in calf earlier.

Cows which are genetically poor milk producers tend to convert nearly all feed to meat and fat, e.g. beef cattle and beefy dairy cattle.

How much should I feed my cow?

Once the weight of the cow is estimated (see table) look up the M/D of the particular feed and then place a straight edge from the cow's weight to the M/D of the feed and look across and you will see

Metabolisable energy required (MJ) to produce 1 kg milk of varying composition

SNF content %	Fat content of milk (%)											
	3.0	3.2	3.4	3.6	3.8	4.0	4.2	4.4	4.6	4.8	5.0	5.2
8.4	4.48	4.61	4.74	4.87	5.00	5.13	5.26	5.39	5.52	5.65	5.79	5.92
8.5	4.51	4.64	4.77	4.90	5.04	5.17	5.30	5.43	5.56	5.69	5.82	5.95
8.6	4.55	4.68	4.81	4.94*	5.07	5.20	5.33	5.46	5.59	5.72	5.85	5.99
8.7	4.58	4.71	4.84	4.98	5.10	5.24	5.37	5.50	5.63	5.76	5.89	6.02
8.8	4.62	4.75	4.88	5.01	5.14	5.27	5.40	5.53	5.66	5.79	5.92	6.05
8.9	4.65	4.78	4.91	5.04	5.17	5.31†	5.44	5.57	5.70	5.83	5.96	6.09
9.0	4.69	4.82	4.95	5.08	5.21	5.34	5.47	5.60	5.73	5.86	5.99	6.12
9.1	4.72	4.85	4.98	5.11	5.24	5.37	5.51	5.64	5.77	5.90	6.03	6.16
9.2	4.76	4.89	5.02	5.15	5.28	5.41	5.54	5.67	5.80	5.93	6.06	5.19
9.3	4.79	4.92	5.05	5.18	5.31	5.44	5.57	5.71	5.84	5.97	6.10	6.23
9.4	4.82	4.96	5.09	5.22	5.35	5.48	5.61	5.74	5.87	6.00	6.13	6.26
9.5	4.86	4.99	5.12	5.25	5.38	5.51	5.64	5.77	5.91	6.04	6.17	6.30

Reprinted from MAFF Technical Bulletin 33 with permission of the Controller of Her Majesty's Stationery Office

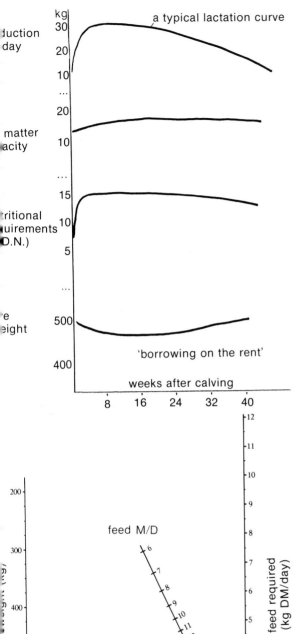

a typical lactation curve

'borrowing on the rent'

weeks after calving

Cattle maintenance requirements

feed M/D

feed required (kg DM/day)

Heart Girth		Est. L.W.		Heart Girth		Est. L.W.	
inches	cm	lb	kg.	inches	m	lb	kg.
25	63	76	35	61	155	668	302
26	66	80	36	62	157	700	316
27	69	84	38	63	160	732	330
28	71	89	40	64	162	766	348
29	74	95	42	65	165	800	363
30	76	101	45	66	167	835	379
31	79	108	49	67	170	871	395
32	81	118	53	68	173	908	412
33	84	128	58	69	175	947	430
34	86	138	63	70	178	987	448
35	89	148	67	71	180	1027	466
36	91	158	72	72	183	1069	485
37	94	168	76	73	185	1111	504
38	96	180	82	74	188	1153	523
39	99	192	87	75	190	1197	543
40	101	208	94	76	193	1241	563
41	104	224	102	77	195	1285	583
42	106	240	109	78	198	1331	604
43	109	257	117	79	200	1377	625
44	112	275	126	80	203	1423	646
45	114	294	133	81	205	1469	667
46	117	314	143	82	208	1515	688
47	120	334	152	83	210	1561	699
48	122	354	162	84	213	1607	730
49	125	374	170	85	215	1653	751
50	127	394	179	86	218	1699	772
51	130	414	188	87	220	1745	792
52	132	434	197	88	223	1791	813
53	135	456	207	89	226	1837	834
54	137	478	217	90	228	1883	855
55	140	501	227	91	231	1929	876
56	142	526	239	92	233	1975	897
57	145	552	251				
58	147	579	263				
59	150	607	276				
60	152	637	289				

An easy method of weighing a small calf is to take the bathroom scales with you, weigh yourself, note the weight, pick up the calf and get back on the scales, note the total weight, subtract your own and you have the answer.

Minerals and vitamins

Milk contains a lot of both so the cow's needs are high: the cow manufactures for her body all vitamins except A and E which she obtains from green grass, green hay and some cereals, e.g. maize. She can store vitamin A in her body for up to six months so vitamins are usually no problem. She needs sunlight in order to produce vitamin D. The B group of vitamins is manufactured in the rumen; however recent American research indicates that cows may not receive sufficient Niacin (one of the B group) from this source at times of heavy production. Most improved pasture mixtures on good soils provide species of plants containing the right *minerals* in roughly the right proportions. However, modern dairy cows

the amount required for maintenance. Add to this what she needs for production and you know approximately how much to feed your cow per day. Of course, it is not usually advisable to offer a feed made up of only one constituant (except say good quality hay or pasture).

Feed	M/D (MJ/kg)	DM (per cent)
Grains:		
Corn (maize)	13.5 (13-14)†	90
Sorghum	13	90
Wheat	13 (12.5-13.5)	90
Oats	12.5 (11-13)	90
Barley	13 (12.5-13)	90
Grain by-products:		
Sheep nuts	11 (9-13)	90
Wheat pollard	11	90
Wheat bran	10	90
Oat bran	9	90
Rice bran	11	90
Hays and grasses:		
Lucerne hay	8.5 (8-9.8)	90
Clover hay	9 (8.3-10.9)	85
Pasture hay (mostly clover)	8.3	85
Pasture hay (mostly grasses)	8.3	85
Phalaris hay	8.2 (7.8-8.7)	85
Oaten hay	9.3 (8.5-9.5)	90
Wheaten hay	8	90
Oat, barley or wheat straw	5 (4.5-5.5)	90
Mature sown grasses	7.2 (6-8)	90
Mature native grasses	6.8 (5.8-7.9)	90
Crop residues∗*:*		
Wheat stubble	5.1 (4.8-8.2)	90
Barley stubble	5.5 (5.1-6.2)	90
Rice stubble	5.7 (5.3-6.6)	90
Oat stubble	4.6	90
Sorghum stubble	8	90
Green feeds:		
Pasture (mixed grasses and clover, succulent and closely grazed)	12	20
Pasture (mixed grass and clover, flowering)	10.3	25
Perennial rye grass	8.3	25
Clovers	9.2 (8.7-9.7)	20
Lucerne (in flower)	8.3	25
Oats (feeding-off stage)	9.3	20
Oats (in flower)	8.7	25
Wheat (feeding-off stage)	9.3	20
Maize	8.7	20
Sorghum	8.5	20
Phalaris (closely grazed, succulent)	8.4	25
Silages:		
Pasture (mostly clover)	8.6	20
Mixed grass and clover	8.2	20
Maize	7.5	25
Oat	8.3	30
Lucerne	8.4	25
Wheat	8.7	30
Protein meals:		
(per cent crude protein)		
Meat (40-55)	11	90
Meat and bone (40)	10	90
Fish (55)	11	90
Peanut (42)	11	90
Cottonseed (40)	10.5	90
Linseed (30-35)	11.5	90
Coconut (20)	12.5	90
Sunflower (40-45)	10.5	90
Safflower (45)	11	90
Soybean (50)	12	90
Lucerne nuts (20)	9.5	90

Miscellaneous:		
Citrus pulp	13	10
Molasses	13	75
Pumpkins	12.5	10
Turnips	12.5	10
Potatoes	12.5	20
Black oats	9	90
Poultry litter	8.5	85
Willow leaves and fine stalks	9	30
Brewers grains (dry)	9.5	90

sometimes produce so much milk they cannot get enough grass into the space inside them in order to get all the energy, minerals and protein they need and hence rations have to be supplemented.

The main minerals needed are calcium, phosphorus and occasionally salt (NaCl). A Ca:P ratio of 2:1 is a good rule of thumb. Tricalcium phosphate or steamed bone meal are both pretty close to this and are usually added at the rate of 1% of a grain mixture as is common salt.

Is this knowledge necessary?

As cattle in peasant and nomad cultures have survived (though often miserably) for thousands of years, one might wonder if all this heavy stuff about megajoules is really necessary.

Well, it is helpful for many reasons, though by no means essential. Just by observing:

a) the cow's condition,
b) her behaviour,
c) the state of the pasture, and
d) the quantity of milk she gives, you will be guided as to her nutritional requirements.

However, these factors must be looked at together and in balance. A cow, like most other domestic animals, can be quite adept at telling a very sad tale about all the grass in her paddock being quite inedible and that she is about to expire if she doesn't soon get a few kgs of pellets. True, cows often moo when they are hungry but they may be just 'nice hungry'. If they are in reasonable condition and giving sufficient milk, then turn a blind eye and a deaf ear to their importuning.

If, however, the four parameters above indicate a feed deficiency, a couple of biscuits of hay a day and a few litres of concentrates will certainly help. Now, returning to 'science'.

Nutritive values of some feeds

feedstuffs	dry matter %	crude protein %	digestible protein %	total digestible nutritients %	starch equivalent %	calcium (grams per lb)	phosphorus (grams per lb)
Low protein dry roughages:							
Oaten, barley, wheaten or sorghum straw	90	3.0	0.5	40	15	0.9	0.45
Oaten, wheaten, sudan grass or Jap millet hay	5.5	3.0	50	40	0.9	0.9	
Pasture hay	90	6.0	1.0	45	35	—	—
Peanut straw	90	6-10	2.0-5.0	5	35	—	—
Corn stalks	83	—	0.8	41	35	1.48	1.00
High protein dry roughages:							
Lucerne hay	90	15	10	50	40	5.4	0.9
Cowpea, field pea or soybean hay	90	16	10	50	38	5.4	1.4
Peanut hay	92	13	10	55-65	45-55	5.1	0.67
Low protein succulent roughages:							
Green fodder or silage from summer crops	25-30	1.5	1.2	18	14	0.4	0.4
Green fodder or silage from winter crops cut at flowering	25-30	2.0	1.5	17	12	0.4	0.4
Sugar cane	25	1.0-3.0	0.6-1.5	15	10	—	0.18
High protein succulent roughages:							
Lucerne	25	4.5	3.5	14	12	1.6	0.3
Cowpea or field pea	20	3.0	2.0	12	10	1.1	0.2
Young grazing crops (growing stage)	25	3.0	2.0	14	12	0.4	0.3
Grains:							
Maize	90	10	8	80	78	0.05	1.4
Grain sorghum	90	10	8	80	75	0.1	1.4
Wheat	90	11	8	78	72	0.2	1.8
Barley	90	9	7	77	71	0.2	1:6
Oats	90	10	8	72	60	0.3	0.9
Grain by-products:							
Wheat bran	90	15	11	63	55	0.4	4.5
Wheat pollard	90	15	11	73	65	0.4	3.0
Protein rich concentrates:							
Meat meal	90	55	45	78	75	26.0	14.0
Meat & bone meal	90	45	30-40	65-70	60	37.0	17.0
Peanut meal	90	42	37	80	73	0.70	2.7
Cottonseed meal (decorticated)	90	41	33	75	70	0.90	5.4
Linseed meal	90	32	26	75	65	1.80	3.4
Coconut meal	90	21	17	80	74	1.40	3.0
Urea (46% nitrogen)	90	equivalent to 260	—	—	—	—	—
Miscellaneous:							
Cottonseed hulls	90	3.9	—	44	35	0.59	0.27
Peanut hulls	92	3.3	1.6	19	10	1.10	0.27
Pineapple pulp	15	2.0	0.13	10	8	—	—
Molasses	75	3.5	—	60	50	4.5	0.4
Sheep or cattle nuts	—	12-18	9-14	75	65	—	—

Sources of nutrition

The next step after looking at *what* a cow needs is *where* it can be found.

As mentioned above high class pasture is normally an ideal feed for dairy cows as it contains all essential nutrients but not in a sufficiently concentrated form for very heavy milkers.

House cow owners do not usually want heavy production, therefore supplementary feeding is only necessary when pasture is scarce or of poor quality.

Cows can also be conditioned to eat a wide range of household scraps (cabbage and cauliflower leaves, orange, apple and potato peel, stale bread, pumpkin and melon skins, and so on). Anything of animal origin is usually more difficult to introduce into the diet, but a little of it hidden in familiar food often works. Beware of poisonous plants such as oleander, young or stunted sorghums, red lantana, bracken fern, arum lillies, azaelas and others. Cows usually avoid toxic plants unless pressed by adversity. However, be very careful when feeding lawn clippings where species are all mixed in as these could contain oleander leaves which are deadly even if old and dry.

Some animals develop habits such as chewing up plastic sheeting and bags, rubber gloves, dirt and old bones. While the rumen is a large organ, space taken up in it by paraphenalia of no feed value and which is slow to break down and pass through, reduces the rate of feed intake. The reticulo-rumen contents of slaughtered animals also often reveal a wide array of pebbles, nuts and bolts, nails and wire. The latter two can be fatal if they penetrate the reticulum wall and enter the heart.

The rule is therefore to keep paddocks free of rubbish, particularly bits of wire and nails and insert a gastro-magnet into the rumen if you are in doubt or the animal is valuable.

Calculating the relative cost of feeds

The house cow owner should aim to feed his cow from pasture all the year round.

One way to help achieve this is to calve the cow in the early spring and dry her off early in winter. This allows the milker the two coldest months for 'sleeping-in' and reduces the cow's feed requirements when feed is usually in shortest supply.

However, it is sometimes necessary to buy in concentrate and/or hay.

To do this economically one should, with certain qualifications, take the obvious 'super-market'

approach and buy that which is cheapest on a 'unit value' basis, i.e. the feed with the cheapest units of energy and protein. It is just like buying washing up detergent — the cheapest by volume is not always the most economical.

Energy and protein are the usual nutrients we are after.

Once we have an idea of the amount of these a food contains (see table page 44) it is not difficult to work out a cost per unit. However, confusion arises due to the variety of methods of measuring energy for example.

Older books may quote feed constituents and animal requirements in terms like Starch Equivalent (SE), Total Digestible Nutrients (TDN), and the measurement may be by weight or percent or calories rather than megajoules.

The unit chosen does not really matter that much; however, Departments of Agriculture in Australia and in a number of overseas countries have adopted the ME (Metabolisable Energy) system as it has more inbuilt refinements and qualifications should these be required.

Balancing the ration

As cows require about 10% protein for maintenance and 20% for milk production, milking cows are normally fed about 15% protein in their total ration, therefore, the farmer must note any grazing as to its quality and quantity.

By quality we mean megajoules/kg dry matter (M/D) and protein percentage. Pasture from half grown to maximum leaf area stages provides the optimum *quality,* and maximum leaf area to flowering for maximum *quantity.*

When cows have unlimited access to pasture we can only roughly estimate what supplementary feed, if any, the cow needs to balance her ration.

Protein in general is more expensive than energy feed so we tend to feed the cow no more protein than is necessary, hence we 'balance the ration' to make sure she has sufficient of both energy, protein and perhaps minerals.

An example

If, say, oats worked out the cheapest energy feed and sunflower meal the cheapest protein, we may then wish to work out say a 15% protein concentrate to supplement pasture of fair quality and availability. Say the oats is 10% protein and the sunflower 40% and we want 15% in the mixture.

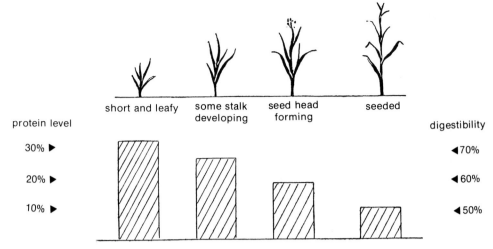

The protein level of pasture drops with maturity. Pasture should be managed to keep it short, green and leafy, mature dry pasture holds its feeding value only so long as it is not subject to rain.

Take 100 kg oats = 10 kg protein
Take 20 kg sunflower = 8 kg protein
Total 120 kg of which 18 is protein

$$\frac{18}{120} \times \frac{100}{1} = 15\% \text{ — a lucky stab.}$$

The above method is called trial and error. Another method is Pearson's Square.

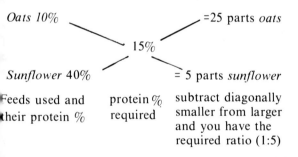

subtract diagonally smaller from larger and you have the required ratio (1:5)

Pastures

For cheap milk production pasture reigns supreme *provided* it contains suitable species at the right stage of growth and is well managed. By management, I refer to grazing pressure, fertility and moisture status and presence of weeds.

Temperate species such as ryegrass, cocksfoot, and clovers are usually superior in digestability and palatability to those species that grow best in warm weather such as kikuyu, paspalum, rhodes grass and couch. However, the latter are perennial species and often dominate pastures in much of coastal Australia. Their great advantage is that they are permanent and tend to dominate and compete successfully with most weeds because of their dense root systems.

Species to use

District agronomists will be able to give helpful advice for your particular region both as to species best suited to your district and establishment methods.

As a general rule in coastal districts with a subtropical climate (from about Nowra N.S.W. north) paspalum and kikuyu can form the backbone of the pasture. However, they produce little feed in the cooler months. It will be necessary to establish winter growing species in order to provide all year round feed.

Establishment of pastures

There are many possible strategies and the best combination for you will depend on the area size you are attempting to establish, the availability or otherwise of machinery or contractors, your attitude to the use of weedicides and chemical fertilisers, the present condition of the land, the rainfall pattern and the availability of irrigation.

It may often be cheaper to use extra seed and broadcast by hand rather than get a paddock

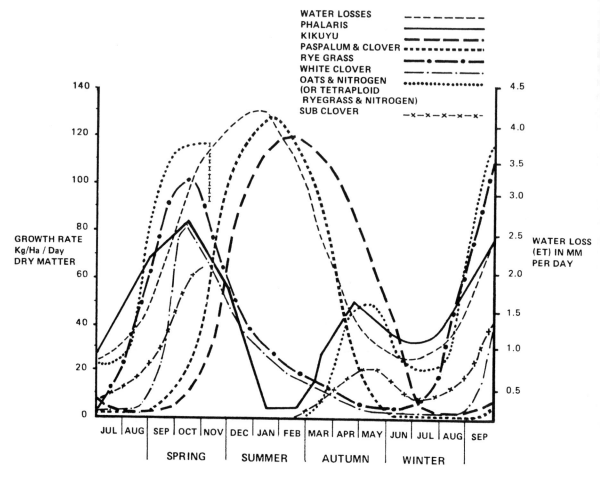

Legend (top right of graph):

WATER LOSSES — – – – – – –
PHALARIS — ————
KIKUYU — ━ ━ ━ ━
PASPALUM & CLOVER — •••••••••
RYE GRASS — ━•━•━•━
WHITE CLOVER — ━·━·━·━
OATS & NITROGEN (OR TETRAPLOID RYEGRASS & NITROGEN) — ••••••••••
SUB CLOVER — –x–x–x–x–x–

Left axis: GROWTH RATE Kg/Ha / Day DRY MATTER — 0, 20, 40, 60, 80, 100, 120, 140

Right axis: WATER LOSS (ET) IN MM PER DAY — 0.5, 1.0, 1.5, 2.0, 2.5, 3.0, 3.5, 4.0, 4.5

Bottom axis: JUL AUG SEP OCT NOV DEC JAN FEB MAR APR MAY JUN JUL AUG SEP

Seasons: SPRING | SUMMER | AUTUMN | WINTER

The above graphs indicate growth rates per day and apply to good seasonal conditions or irrigation. The kikiyu has nitrogen top-dressing in autumn only.

In this graph, the phalaris production is abnormally high in winter and would usually be about the same as oats. Water loss is from evaporation from the soil and transpiration from the leaves.

cultivated or use a fertiliser spreader or pasture seed drill.

Most temperate species establish fairly easily if broadcast in mid-autumn with superphosphate. Fertility is of great importance in establishing cool season species and superphosphate (or Mo super) is one of the keys to fertility. It is possible to have excellent ryegrass and white clover pastures in the winter and spring covering an area established to kikuyu and paspalum. Legume seed (clovers and lucerne) may need innoculation and lime pelleting if being introduced to an area not previously supporting these species. Packets of innoculum are inexpensive and carry all the instructions.

Management

When some pasture species get too tall (as with kikuyu and paspalum in good seasons) they should be mowed or slashed. If you are short of feed then spread a little nitrogen fertiliser such as Nitram ® during or immediately prior to rain (or water-in) and this will transform a stalky unpalatable patch of kikuyu, but mow or slash the rank growth first.

Poultry manure from broiler or layer sheds is also excellent for encouraging pasture growth, especially on sandy soils.

For the cow owner it should become second

nature to be anticipating the feed situation a week and a month in advance. If your paddock is not that big then heavier fertilisation, optimum species mix, irrigation and weed control can enable a greatly increased quantity of feed to be grown on a small area. In these situations it is essential to have a cow that tethers and leads easily so that during times of feed deficit you can give your paddock a spell while you utilise odd pieces of ground in the neighbourhood.

It is not legal to tether animals along roadsides or in parks unless they are attended. Animals can only be agisted along roads if they are attended. A permit for mobs or flocks must be obtained from a pastures protection board. However, in practice, whilst a council impounding officer can impound 'unattended' animals, he usually uses his discretion as to what interpretation to put on the word 'attended' in the particular circumstances and whether any complaints have been made.

How much pasture area is necessary to support one milking cow?

Any answer is of course, subject to many qualifications such as:

a) Mean daily winter and summer temperatures
b) Total rainfall and its reliability and distribution
c) Evaporation rates
d) Soil fertility
e) Physical structure of soil (affects infiltration rate of rainfall or irrigation)
f) Pasture species present
g) Density of pasture sward
h) Size and producing ability of the cow
i) Quality of grazing management

One acre of reasonably fertile, well-drained soil carrying dense pasture of mixed temperate species, in a mild climate, with high and dependable rainfall (or supplementary irrigation), with only low to moderate evaporation rates and with clever and careful management and weed control, and regular fertiliser application would easily support one small to average sized house cow provided she was dry during the two coldest months.

As less favourable conditions apply, increased land areas or more supplementary feeding will be necessary if milk production is to be maintained.

Tethering your cow on odd patches of unused land, quiet roadsides and even your own garden, can greatly reduce the pressure on your paddock at times of increased pasture demand or slow pasture growth. If your cow tends to eat your favourite shrubs and plants (and most cows cannot be trusted) then placing old blankets or canvas over or around them and tying up or weighting them down will often allow the cow to graze right up to the edge of the plants but leave them unharmed.

Fencing and watering

If you need to erect a considerable amount of new fencing (over a kilometre), electric fencing should be considered. It is much easier to erect and move about. Three strands of barbed wire as a standard fence will keep most cows in if it is strained tight, but not necessarily young stock who will not only make a good attempt at getting out but tend to cut themselves about.

The type of fencing will also be determined by what stock or pests, if any, may be on the other side of the fence.

Cows drink a lot of water — about a litre for every 10 kg bodyweight plus 15 litres for every 5 litres of milk produced. This amounts to about 50 litres or just over 10 gals per day for the average mature Jersey cow under mild environmental conditions.

Apart from permanent creeks and dams, an old bathtub makes an ideal trough. If you are likely to be away for extended periods and have no neighbourly arrangements to check stock water, fairly cheap automatic devices are now available which are much more resistant to damage and breakdown than the old toilet cistern variety of ball-cock.

Pastures for South-Eastern Australia

What to sow	*When to sow*	*Fertiliser*	*Remarks*
Dryland Pasture — mixture 6 kg/ha Wimmera ryegrass 6 kg/ha Kangaroo Valley early ryegrass 2 kg/ha white clover (Haifa) 3 kg/ha sub clover (Seaton Park, Woogenellup)	March — May	Sow with Mo super at 250-375 kg/ha. Topdress with Nitram to boost ryegrass growth. Annual topdressing of 250 kg/ha single super.	*Thorough preparation of good seed bed most important. *Allow ryegrasses and sub clover to seed down after flowering. Don't graze too heavily in October. *Broadcast paspalum seed at 3 kg/ha in October to encourage natural regeneration of paspalum
Phalaris — dryland 3 kg/ha phalaris (Sirosa, Sirolan) 2 kg/ha white clover (Haifa) 3 kg/ha sub clover (Seaton Park, Woogenellup)	As above	As above for medium to high fertility soils	*Phalaris is slow to establish — graze lightly in first year *Thorough preparation of good seed bed most important *Broadcast Wimmera ryegrass 1 week after the establishment of phalaris seedlings
Kikuyu — irrigated or dryland 5 kg/ha kikuyu (Whittet or in cooler climates — Croft) 2 kg/ha white clover (Haifa) *or* 4 kg/ha sub clover (Clare) (Sub clover acts as a forage crop in the first year.)	Mid February to mid March	Sow with Mo super at 250-375 kg/ha. Topdress annually with 250-375 kg/ha Single Super and topdress with Nitram as required	*Good seed bed essential *Excellent clover growth in autumn and early s *Graze heavily and strategically to encourage ki root development

ANNUAL WINTER FORAGE CROPS

Oats — Cassia, Saia, Algerian, Blackbutt, Feb-April Saia Irrigated — 100 kg/ha Dryland — 80 kg/ha	Feb-April	Irrigated — 250 kg/ha Single Super 125 kg/ha Nitram *or* 250 kg/ha Starter 18 Dryland — 250 kg/ha Single Super Topdress with Nitram	*Sowing early will ensure good growth before w *Early grazing ensures good tiller development *Allow sufficient regrowth before June for win feed
Ryegrass — Tetila, Tama, Wimmera 25 kg/ha	Feb-April	As above	*Produces great amounts of feed if sown early *Responds to frequent applications of Nitram

ANNUAL SUMMER FORAGE CROPS

Forage Sorghums — Speedfeed, ST6, Cowchow, Zulu, Forage	Mid October	As for oats plus 62 kg/ha muriate of potash	*Allow crop to be 1m high before grazing. W stock if grazed too early due to high prussic a content.
Pearl Millet — Katherine, Tamworth	Mid October	As above	*Graze frequently at 15cm
Japanese Millet	Early Sept	As above	*Graze at 15cm 3 weeks after sowing and then the crop to seed before next grazing.

Dentition of the cow

A calf has 20 milk teeth — There are 8 lower front incisors (nippers) 6 back or grinders in the upper jaw and 6 back or grinders in the lower jaw. There are no upper front teeth but a tough dental pad. All teeth are temporary — 8 incisors and 12 molars.

In the *mature cow* there are 32 permanent teeth. There are 8 lower front incisors, no upper front but a tough dental pad. There are 12 back teeth or grinders in the lower jaw with a corresponding number in the upper jaw.

So there are twice as many molars in the mature cow as in the calf. The comparative timing of eruption of the molars and incisors in cattle is not completely dissimilar to that of humans. The fourth, fifth and sixth molars (permanent) erupt behind the temporary molars 1, 2 and 3. Molars 4, 5 and 6 appear between the ages of 6 and 18 months whereas replacement of molars 1, 2 and 3 occurs between the ages of 2 and 3 yrs 3 mths.

The age of eruption of cattle teeth (particularly incisors) is influenced to a minor extent by inheritance and a marked extent by environment *ie.* as the rate of maturity is generally in livestock.

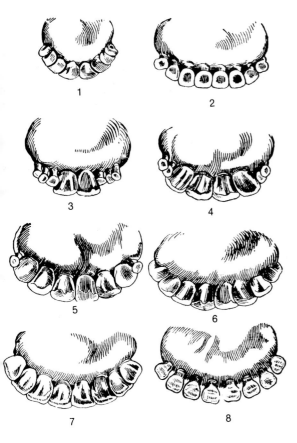

DENTITION OF THE OX, showing development at various ages

1—1 month
2—1 year
3—1 year 10 months
4—2½ years
5—3½ years
6—4½ years
7—4½ years
8—10 years

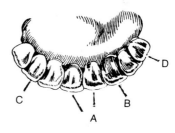

A. central incisors
B. first intermediates (middle or medials)
C. second intermediates (laterals)
D. the corners

Reproduced from the Standard Cyclopaedia of Agriculture

A steer with bloat.

9 Calf Rearing

Care of the cow

A prerequisite for successful calf rearing is the birth of a healthy, vigorous calf. This will only occur if the cow is in good and improving condition as she approaches calving. *Good* quality pasture is sufficient for all stages of pregnancy. If the cow is in poor condition then supplementary feeding to build up her condition is essential if you wish to do your best to ensure she calves easily and has a live and healthy calf.

Care of the calf

A newborn calf seldom needs attention. If left alone the cow usually cleans up the calf and may eat the afterbirth. A calf normally gets to its feet after about 30 minutes and is suckling within an hour. After an assisted birth the calf may need to be helped to commence breathing, to be dried and to be helped to suckle. It is essential that the calf be allowed to suckle its mother for at least a day after calving for two reasons. Firstly the *milk* (known as colostrum) secreted during this period is extremely rich in vitamins (particularly vitamin A) and antibodies (immuno globulins) and secondly the *intestinal walls* of the newborn calf allow for the passage of protein without digestion for only about the first day. After that the immuno globulins are broken down in the gut and lose their identity and disease fighting properties.

Vitamin A is 10 times more abundant in colostrum than in normal milk. Vitamin A is also involved in protection against diseases. Colostrum is also richer in minerals but much lower in lactose than normal milk. Milk replacer high in lactose can cause scours.

If the calf will not stand up or suck, strip the colostrum from the cow and slowly bottle feed the calf using a 'cafeteria' type teat or old milking machine inflation.

Some calves are strong enough, but cannot find the teat (especially with low uddered cows). Others have to be taught to suck. Guide the calf to the cow and squirt some colostrum in his mouth while it is held open with fingers. As he commences to suck, guide him onto a teat.

A little trouble taken during the first few hours of a calf's life may not only mean the difference between a healthy calf and one that receives a severe and expensive check or setback to health, but very often the difference between a calf and a corpse. The secret of success is to be observant and to be sure that the calf has had a feed within a few hours of birth — its appearance (gut filled) and that of the cow's udder will often tell the story.

Synthetic colostrum is available. It is recommended by the manufacturer for the first few feeds for orphaned calves or very young calves bought in. This product contains an anti-scouring agent (see page 56), but it does not contain antibodies and is therefore inferior to the real thing. It is possible to store colostrum either in a fermented form or frozen though the viability of the antibodies slowly declines.

Electrolyte replacers (see page 56) are often given for one feed to calves bought in.

Two or three raw eggs made into a sort of egg-flip can also be a great help to very young calves which have had a poor start. One egg beaten into 1 litre of milk replacer and a pint of warm water, and a dessertspoonful of olive or castor oil can be most useful. It should be fed at about 40° C. Eggs can also be given shell and all.

It is often possible to pick up a gallon of fresh colostrum from a commercial dairy. However, this is not usually necessary because even if the cow dies or is down with milk fever it is mostly possible to either strip out or get the calf to suckle sufficient colostrum to ensure its survival.

When all else fails and no colostrum is available from any source it is advisable to feed the newborn calf 250 mg of auriomycin/day for the first 5 days and 120 mg/day for the following 5 days.

If the calf is left on the cow for more than a day excess production should be removed from the udder each day. This will reduce the risk of mastitis infection or of the cow drying off prematurely, especially if the calf is not suckling all four teats.

If all the cow's milk is required for household purposes, or for some other reason a cow is not available to rear the calf, then it must be reared on a commercially available calf milk replacer. A good replacer consists mainly of cool - spray-dried skim milk powder and homogenised tallow at 15 to 20%.

Following initial feeding on colostrum for 2 to 3

days, milk replacer is fed *twice* daily for a further 4 to 5 days at the following rates:

weight of calf	milk replacer/feed	water/feed
45 kg	250 g	2½ litres
35 kg	200 g	2 litres
25 kg	125 g	1½ litres

Depending on the actual condition of the calf it can now be fed *once* daily until 5 weeks of age at the following rates:

weight of calf	milk replacer/feed	water/feed
55 kg	700 g	3½ litres
45 kf	600 g	3½ litres
35 kg	500 g	3 litres

Note that the amount of water used in once-a-day feeding is about ½ that used in twice-a-day. Calf milk replacer mixes readily with water that has been warmed to 38°C. Calves can be taught to drink directly from a bucket or from a calf nipple. Nipple feeders are ideal where a number of calves are being reared.

It is advisable to tether calves until they are drinking freely. This is especially the case where calves are being taught to drink from a bucket.

Develop the rumen quickly

We have probably all seen show beef cattle that have been fed milk up to an advanced age. While some calves wean themselves, a lot will go on drinking milk as long as it is offered. Their attitude may be why eat vegetables when you can have your fill of dairy foods?!

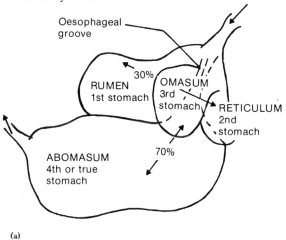

(a)

a) Stomach system of a calf (note undeveloped rumen). Early in a calf's life liquid feed passes directly into the fourth stomach. After two weeks the rumen (first stomach) starts to develop and the calf begins to obtain nutrients from pasture and other fibrous feeds.

As top quality pellets are far cheaper than milk replacers, much thought has been given in recent times to early weaning and speeding up the rate of development of ruminant function in the calf.

Digestion in new-born calves is different to that in adult cattle. Unlike adults, young calves do not ruminate. They do have four stomachs, but for some time after birth, digestion takes place only in the fourth one. Because of this, a calf's feed requirements differ from those of the ruminating adult.

After an initial period on milk only, calves gradually become fully fledged ruminants over a period of about twelve weeks. It is important to understand the nutritional requirements and digestive processes of the calf during this development period.

For the first week or so of life the calf obtains practically all its nutrients from a liquid diet. However, even at a few days of age the calf will nibble grass and other fibrous foods. This type of feed passes into the rumen or first stomach, where it accumulates. Bacteria and other micro-organisms, taken in with the feed, soon become active and start to break it down. This is a fermentation process which stimulates growth of the rumen, and this growth continues until the rumen is fully functional.

There is a transition from wholly milk feeding to wholly grass feeding, during which the calf obtains nutrients from both sources.

Early weaning

Once the transition stage is really under way the calf

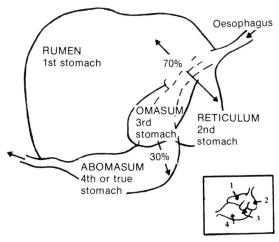

b) Stomach system of an adult cow. Inset shows roughly to scale the size of the calf's stomach system compared with that of an adult.

Agricultural Gazette

can be weaned and it will probably survive. However, a good manager does not allow the survival impulse to take over. Profitable calves are those growing unchecked, not those just surviving.

When calves are fed milk twice a day the savings in labour and feed costs on a per head per week basis of early versus conventional weaning are roughly 100%, i.e. it costs approximately half the amount to rear a calf from 5 weeks to 10 weeks on pellets as it does on milk replacer (with a comparable growth rate). However, for successful early weaning observant and intelligent management is absolutely vital. Especially in watching for scours (calf diarrhoea) and parasites.

How to succeed at early weaning

Use only good quality 18% or 20% protein calf pellets and do not attempt to wean the calf until it is eating about 1½ kg of pellets a day. The best results have been obtained when the calves were given access to straw and water only in addition to the pellets. Straw is preferable to hay as its function is to stimulate the rumen and salivation, not provide nutriment. Hay being more palatable than straw is eaten at the expense of pellets. The result is an unsatisfactory growth rate, because there is not enough nutrients in the hay.

From 12 weeks of age strong and healthy calves will grow satisfactorily on good quality pasture. If this is not available, they should be supplemented with dairy meal or pellets and hay until the pasture supply improves. A growth rate of half a kilo per day is desirable for a young calf. People however do not usually weigh calves but simply note their behaviour, body condition, coat, eye and the presence of any diarrhoea (see page 55 on worms).

The age at which calves can be weaned early will depend on:
1. The age at which a nutritious and palatable pellet is first offered.
2. The comparative birth weight of the calf and whether it has had a check.
3. The ability of the husbandman to gauge when the calf is ready.

When calves are weaned off milk or pellets they will tell a "tale of woe" for a day or so and bellow and bawl frequently and will be very convincing that they are on the point of starvation. However, if you persevere they soon adapt to the new regime.

Rearing calves on cows

The least time-consuming method is to rear the calf on a cow. When milk is required for household use then the calf should be kept apart from the cow during the day. After milking out the cow in the evening the calf is left to suckle the cow till morning. Fostering an additional calf or calves should be considered when it is apparent that the cow's milk supply is in excess of that required.

Where difficulty is experienced in getting a cow to accept a foster calf, one or more of the following methods should be tried:
1. Restraining the cow by haltering and leg roping twice a day for the calf to suckle until it is accepted, or tying the cows hocks together.
2. Tying the calf to the cow by means of a long chain.
3. Tethering the cow's own calf with another calf until the stranger is accepted.
4. Confusing the cow by removing its own calf and replacing it with others which have been smeared with a strong smelling substance such as neatsfoot oil which had been previously placed on the cow's muzzle and own calf.
5. Leaving the cow unmilked for 12 to 24 hours and also keeping the calves to be placed on her unfed for a similar period. Their enthusiasm and her relief often combine successfully. A confined space such as a crush or a very small yard is sometimes used in this case. Some few cows just will not take fosters even after weeks of perseverance. Other cows will take any calf at any time.

The clue is to start by getting the calf 'hooked' on the cow. This is usually more straightforward than getting the cow hooked on the calf.
6. Blindfold the cow for a few days.
7. Introduce a dog to the cow and calf. This will sometimes trigger care-giving behaviour in the cow.
8. Some cows will accept a calf after being given an inter-uterine infusion (this seems to trigger a physiological response similar to that which a cow experiences at calving).

Calf rearing as a sideline

Calf rearing can be an economic proposition if you buy wisely and manage carefully, but it can still be uneconomic in spite of the best management as the calf, dairy herd replacement and beef markets all tend to fluctuate considerably.

Always start such a project on a trial basis. Cut your teeth on a few. You can learn from books and from others, but you will learn most in your own calf pen.

Capital outlay can be kept very small and should consist of little apart from direct costs. Expensive sheds and single pens are probably never warranted apart from a tax concession point of view. Some form of portable nipple feeder is all you need. One novel set-up was an old unregistered utility with a long nipple-trough at the back for backing through the fence and a washing machine on the back for mixing the replacer. The owner would plug-in the machine in the shed and drive to the paddock with the hot mixture ready for the calves.

When buying calves, make sure the calf is a few days old and has a dry navel. Very young calves can be a poor proposition. Sometimes calves are medicated prior to sale and sold with a guarantee. If possible avoid auctions and buy from a dairy or raise on a contract basis (unless you breed your own).

Rearing systems

Whatever the system adopted, remember to *reduce stress* to a minimum by using the important management tool of *conditioning*. Make all changes gradual, particularly in relation to what is fed. *Changes in routine should likewise be gradual* as calves get into set patterns and habits. If the calves are properly trained through a little patience being used over the first few days, they will repay the operator by moving quickly through their feeding routine.

The exception to this conditioning idea is when removing an animal from its mother, or carrying out some surgical operation, when 'the quicker the better' applies. The strategy then is to minimise all other stresses and so dissipate any chance of a cumulative effect.

Kindness and common sense are very important. While a calf's needs are mainly physical, they do have *psychological needs*. A calf will often pine away if ill-treated. This is also one reason why calves on cows almost invariably do better than those reared on the bucket, even if fed whole milk.

Most calfhood troubles are associated with feeding. The calf as an animal is particularly prone to scouring. Even calves running on their mothers can develop white scours.

Artificial rearing

Nipples are usually easier and cheaper to use than buckets if any number of calves are to be fed. Trying to teach some calves to drink out of a bucket can be quite infuriating. Two rules are start early and do not give in. Get the calf sucking your fingers and lower them into the bucket. Standing astride the calf with its head between your legs and its stern backed into a corner is ideal for the first time or two. If the calf will not suck he may not be hungry enough. A 24 hour starve will usually overcome this difficulty.

When the calf commences to suck, force the fingers apart to allow milk to pass up into the mouth. Once the calf masters that idea gradually remove the fingers till one only is just inside his mouth and after a couple of feeds at most try him without any. Usually the calf will try a couple of gulps and then go searching for something to suck. Be firm.

Another great advantage of nipples is their use in automatic milk dispensers and systems where the calf is allowed to ad-lib feed.

A 45 litre (10 gallon) milk can or 200 litre (44 gal) drum fitted with nipples and tubes can be used as ad-lib feeders. Fermented colostrum or acidified milk or milk replacer is used in these. The effect of the acidity is to not only preserve (pickle) the milk but to reduce consumption to a safe level. Acetic acid (technical grade) at 1 ml/litre of milk is usual. Up to 4 days supply can be mixed up at once.

Housing requirements are minimal in much of Australia provided that some shade and shelter is available. Protection against wind and excessive wet and cold conditions is necessary during the first weeks of life.

Diseases of calves

In this country we are fortunate to be free of many of the diseases which cause heavy losses in North America, and we have a climate very suitable for raising calves out of doors.

Our principal troubles are worms, lice, and scours. Pneumonia as a secondary infection may put the final nail in the coffin. Calves already checked and in poor condition from scours are often more prone to succumb to worm and lice infestation.

Most other infectious diseases important in Australia can be controlled by vaccination.

a) **Vaccinations** Most of populated S.E. Australia is now free or provisionally free of Brucellosis and so calves in these areas are no longer vaccinated against this disease. Check with your local government vet if you are uncertain.

However, it is advisable to vaccinate calves with '5 in 1' against clostridial infections such as tetanus, blackleg and entero toxaemia (pulpy kidney). Most farmers do this themselves. If you are rearing calves as a sideline, it *may* be worth your while to buy the vaccine and inject them yourself. However, it may be easier to get the vet or some neighbour or A.I. technician who vaccinates as a sideline to do your calves. They will need their first shot at 2-3 months and at least one booster.

b) **Worms** As with vaccinations, the decision

whether to treat your own calves will depend on numbers. The odd calf raised in relative isolation and kept growing may never need worm treatment, while those raised with other calves in the same old yard year after year are sure to be well infested. Paddock rotation is very important in parasite control. Ostertagia or small brown stomach worm is the principal culprit in S.E. Australia. It is most active in the cooler months. A strategic drench in autumn and another in spring may be necessary. Adult cattle are fairly resistant to worms.

Symptoms of worms may include general depression and listlessness, sunken eyes, fluid under the jaw, scours and a dull coat. Hopefully your calves will never get to this stage.

Worms can be treated with powders mixed with the feed, medicated lick blocks, drenches, subcutaneous injections (see p. 62 health), 'pour-on' anthelmintics or injections directly into the rumen using a special gun.

c) **Scours** Any calf will scour if feeding management is faulty. Too much or too little milk, too great a variation in volume, temperature or formulation can all trigger scours. Such scours are often termed 'dietary' in that the causal organism, E. coli, naturally occurs in the lower gut and only invades the upper gut under specific dietary regimes. Some strains of coli are more virulent than others. The faeces are loose and grey to white in appearance. If the calf is allowed to continue scouring it will become dehydrated and may die due to loss of body salts. Treatment techniques for dietary scours usually aim at:

1. reducing gut activity by the temporary cessation of feeding;

2. supplying fluids and electrolytes to replace those lost. The principal electrolytes lost in scouring are sodium, potassium, chlorine, phosphorus and bicarbonate ions. A teaspoon each of salt, muriate of potash (fertiliser) and bicarbonate of soda in 2-3 litres of warm water is a good home remedy. The better commercial electrolyte replacers also contain a 'rehydrator' or mixture of glucose, glycine and citrate which stimulates the cells of the gut lining to absorb larger quantities of fluids. Gradually re-introduce milk after 1-2 days on electrolyte replacer.

Calf scours can also be caused by pathogenic bacteria such as salmonella or coccidia. In such cases the scours may be grey or green and often contain blood and hence are known as 'blood scours'. Treat as for 'white' or dietary scours but isolate infected calves. Treat orally with antibiotics in the case of salmonella and sulpha drugs for coccidiosis. An intra muscular injection of streptomycin and penicillin may also be necessary. A

number of oral anti-scour tablets are available commercially.

d) **Lice and bush ticks** Heavy lice infestations can occur within a few weeks of birth, especially in cool weather. Carefully examine the coat especially between the back legs, around the face and under the neck.

Bush ticks can cause calf losses on parts of the North Coast of N.S.W. in late winter and early spring. 'Pour-on' or 'spot-on' licicides are usually effective. A repeat treatment may be necessary three weeks later.

e) **Pneumonia** While usually a secondary infection, it can be caused by milk or drench incorrectly administered and ending up in the respiratory tract. Treat with antibiotics and keep warm and dry.

f) **Navel infections** Keep a close eye on the navel of young calves and never cut the cord of a new born calf close to its body. If the calf has been dropped in unsanitary conditions, swabbing the navel with iodine immediately after birth might help. A strain of coli bacteria sometimes gains entry to the body via the navel in new born calves and may lead to swollen and infected joints, liver abscesses and death. Repeated I/M injections (p.62) with a long-acting antiobiotic can cure.

g) **Warts** These usually disappear in time of their own accord, ringworm likewise; neither responds dramatically to treatment.

h) **Pink eye** This is a bacterial infection transferred by flies. If caught early (when the eye has just started to weep) it responds readily to a number of antibiotics applied to the eye in the form of a powder (puffer pack), pressure pack spray or ointment. A small squirt of mastitis antibiotic is quite effective.

Management practices in calf rearing

Dehorning Dehorning causes only minor discomfort to the calf and prevents possible serious injury to both man and animals at later stages in life. The horn grows from the skin, the skull (frontal sinus) later grows out into the hollow interior.

The idea is to remove the horn forming tissue. It is usually possible to feel the horn buds at birth or within a week. Once you are sure you have located them dehorning can be carried out.

Two methods are available that can be applied to calves of a few days of age up to about one month.

1. Hot iron method A rounded hollow hot iron kills

he horn forming tissue. There is no blood lost due to the cauterising action of the hot iron. An electric soldering iron where the 'iron' has been replaced by a piece of hollow copper pipe is ideal, or an old fashioned soldering iron will do when the end has been chopped off and a 1½ to 2 cm hole bored in it (the older the calves the larger the size).

Clip the hair around the bud and restrain the calf very firmly on the ground (two people). Apply the hot iron over the bud, press hard, rotate the iron and the angle. On completion there should be a 6 mm deep cut straight through the skin right around the horn bud effectively isolating it from its blood supply.

2. **Chemical method** A number of commercial dehorning pastes are available. Clip the hair as before and apply only the recommended amount of paste. Precautions: do not leave the calf outside if it looks like rain within 12 hours; isolate calves from one another, and smear vaseline around the horn bud before applying so as to confine the paste.

A sharp pen knife can be used to remove the horn bud from young calves (much like taking the stalk and out of a whole apple). Scoop dehorners can be used on older calves. However, hot iron dehorning is the most quick and satisfactory method. Electric hot-iron dehorners at present (1983) cost around $17.35.

Removing extra teats

Female dairy calves often have teats additional to the four normal ones. These teats may not be particularly harmful but they do detract from the appearance of the udder and sometimes secrete milk. Extra teats should be removed while the heifer is still small and easy to handle; the best age being 4 to 6 weeks.

Apply iodine around the teat to be removed. Pull the teat down and snip it off cleanly where it joins the udder. Use sharp, clean scissors and cut off teat at first attempt.

In the case of older heifers extra teats can be removed by placing an elastrator ring around it. The teat drops off in about 10 days leaving a completely healed area.

Castration

Unless beef prices are high and/or the calf is a beef cross or Friesian it may be more economical to sell it as a bobby unless you want to keep it as a fill-in 'milker' when you're away. Keeping bull calves for

beef is also fine for those who do not become emotionally involved with the animals they rear. In these circumstances it is only necessary to castrate if you plan to keep the calf more than 9-12 months.

It is not always easy to find the testes in very young calves as they may have drawn them back into the abdomen.

Once they are both held firmly in the hand castration may be effected by means of a bloodless emasculator which crushes the spermatic cord and blood vessels whilst leaving the skin intact, an elastrator which achieves the same purpose over a period of time or various surgical options. The latter may include a sharp knife or scalpel. With bulls over 15 months seek the advice of an experienced person or vet.

When surgically removing testes it is wise to clean the scrotum well with iodine or methylated spirits. Hold the scrotum above the testes and make a cut or slash right across the base of the scrotum from front to back or vice versa, slicing the bottom half of the teste in half. The teste will then pop out. Scrape through the cord and blood vessels in order to cause maximum damage to the arteries and so minimise bleeding.

If you fail to make a decent sized cut you will be poking around interminably (or it will seem so to the bull) trying to cut through the inner sack and make a hole large enough to get the teste out. Repeat for the other side of the scrotum. If this whole thing is not your scene an experienced neighbour, A.I. technician or vet could help.

Identification and branding

Branding legislation varies amongst States. In New South Wales (1989) it is currently under review, the plan being to re-register all brands and thus weed out those no longer in use. All new applications will be through local pastures protection boards and new brands will carry a district identifier. Positions for registered brands will be left rump, right rump, left hip, right hip etc., and freeze branding will be included as an alternative to fire branding. It is not compulsory to register a brand or brand cattle. It is only necessary if there is a risk of 'lost, stolen or strayed'. Most offices of departments of agriculture have information pertaining to stock branding and will be able to advise you.

Do not brand very young animals (under 6 months) because their skin is too thin, the areas for branding too small and the brands will grow too large as the animal grows.

Before branding clip the coat if very woolly and heat the iron to a dull red. The animal must be well restrained. The pressure applied and duration of the branding operation will depend on the heat of the brand and the age of the calf. The resultant brand should have a bronzy appearance. Always apply the brand at right angles to the skin and be prepared to move with the animal or the iron may slip and smear the brand.

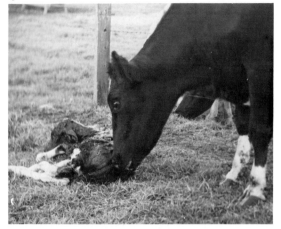

Leave the cow alone as long as she has taken to her calf and is cleaning it up.

Calves naturally orientate to the underline of the cow and will locate the teats providing the udder is not too low.

The typical look of a wormy and scoury calf.

1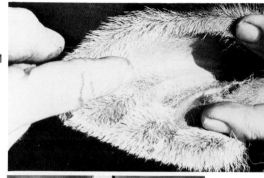

2

3

4

Tatooing a calf 1. The correct spot to tattoo, place the tattoo towards the top of the ear and between the cartilaginous ridges. 2. Tattoo the ear from the front and the top. Test it out on a piece of cardboard beforehand. 3. Rub on the tattoo ink or paste. 4. Squeeze home the tattoo. 5. Rub in excess ink, wipe off and the job is done.

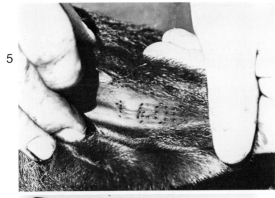

Tethering 1. For cows, have a length of about a metre of chain attached to the collar and then have a clip so that when milking it is very easy to clip up to a fence or tree. The type of clip illustrated is the most satisfactory.

2. Calves can be tethered to a round piece of concrete which can be rolled about.

3. A chain on a wire is another method of tethering.

Putting a calf on the ground 1. Grasp the calf at the dewlap and flank and lift.
2. Push the calf away with your thighs and roll her onto her side.
3. The calf is down. Move quickly to restrain her. Note the automatic calf feeder in the wall (background, right).
4. With the calf on her left side place your right foot between her thighs (uppermost leg to rear); place your right knee into her flank, curl the lower front leg up and back and place left hand on her neck.

A cheap self-feeder for pellets — an old metal garbage can.

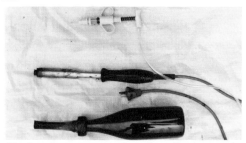

Displayed are: top — *a cheap but effective vaccinator;* middle — *an electric (hot iron) dehorner;* bottom — *an effective 'drenching gun' made from a beer bottle and an old milking machine inflation.*

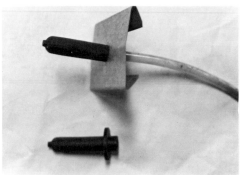

Rubber calf nipple or teat. The uppermost one is fitted for feeding through a wire fence with the galvanised iron bent so the wires sit in the inside angles.

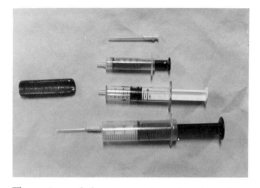

Three sizes of throw away hypodermic syringes — 5ml, 10ml and 25ml. The needles come in a separate pack and are covered by a plastic sheath. To the left of the syringes is a gastro magnet.

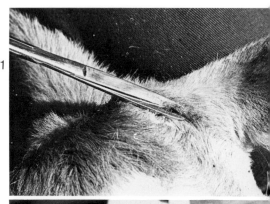

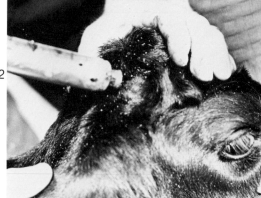

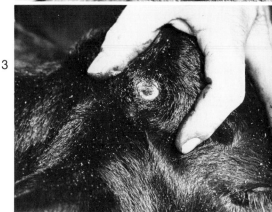

Dehorning with a hot iron *1. Locate the horn bud and clip away hair.*
2. Apply dehorner to exposed bud.
3. The finished job.

Freeze branding, ear tagging, photography and ear tattoos may all be used by stock managers to aid their record keeping. One of the latter two may also be required by breed societies for registration purposes.

10 Maintenance of Health

Cattle, especially crossbreds, are probably the healthiest and toughest of all domestic animals.

Government animal disease and quarantine regulations have done much to not only keep Australia free of some of the worst pests and diseases of cattle but have led to the reduction or elimination of some important pests and diseases accidentally introduced in the earlier years of the colony. All owners of stock should contact the local Pastures Protection Board and be supplied with a tail tag number.

Fundamental to maintenance of stock health are the powers of observation of those managing the stock. To be able to notice things that are not quite right and catch a problem in its early stages is so often the key to success. Another key is to be alert to what can possibly go wrong and take preventative action.

This book is in no way intended as a book on veterinary medicine.

The following short notes, and references to diseases the author has made elsewhere under the headings of calf rearing, reproduction and feeding are included in order to heighten awareness of the cow owner to the possible health problems which might arise and that he may make the best decisions in recognising abnormalities, the degree of seriousness he might attach to them, and what line of action he might pursue.

Diagnosis of illness

Once having arrived at a diagnosis, i.e. whether the animal is in pain, is suffering from a physical injury, infection, metabolic disorder, poison, chronic nutritional deficiency, photosensitisation (sort of acute sunburn), etc. it is then possible to reach a decision about treatment.

Today there is a formidable array of highly effective drugs and antibiotics and other chemicals which when correctly prescribed and administered, can bring about a dramatic reduction in symptoms.

Readers will vary greatly in their educational background and interest in animal anatomy, physiology and pharmacology.

For those who wish to venture someway along the path of home treatment, the following notes may be of some help. Remember that veterinary books are available which contain differential diagnosis charts in order to help farmers zero in on the causal problem. Good powers of observation are the key to problem awareness. Correct diagnosis is the key to success and the key to diagnosis is asking the right questions.

Experience and knowledge help with knowing what possibilities to dismiss and what to follow up. Cardinal signs such as body temperature, rumen motility (periodic contractions), a wet nose, breathing, etc. all help.

Even if the animal is dead it may be worth doing a post mortem if you are not sure of the cause of death as it may be a disease which could spread to other animals. A government vet may give you helpful advice over the phone and should be contacted where sudden deaths have occurred.

Treating sick animals and administering drugs

This type of book is not usually the place for telling funny stories, but the following one I hope has a lesson for us all.

Arriving at the dairy one morning we found our favourite dairy cat cold, wet and shaking uncontrollably. But that was not all. As he walked along the path he jumped about a foot high as he crossed a crack in the concrete and he followed this in quick succession with a jump about two feet high when he saw an ant.

Highly amused at our hallucinating cat we were wondering as to the cause when we noticed the saucer which had sat on the window-sill full of fly bait now broken on the ground and licked clean. The fly bait label said to administer atropene as an antidote. We had a bottle of atropene tablets so we opened Tom's mouth and gave him one. Not long after we noticed our patient rather preocupied with all the bubbles coming out of his mouth and nostrils. He looked rather like a pile of snails after they have been sprinkled with salt —all froth. We then took another look at the atropene bottle and it said 'injectable'. Thankfully Tom survived the treatment which was probably worse than the complaint — he had 'suffered at the hands of many physicians'.

Where to give injections

They may be given:

1. Under the skin, i.e. *sub-cutaneous* (S/C). Loose skin behind the shoulder and elbow are ideal sites, or the neck.

2. Into the muscle — *intra-muscular* (I/M). The rump is the favourite spot with stock over 12 months and the back of the thigh for calves. A couple of smacks with the back of the hand first and animals do not seem to feel the needle so much when it goes in (about 30-40 mm). Put the needle in first and then attach the syringe. This will reduce the risk of breaking the syringe and spilling the contents.

3. *Intravenously* (I/V) usually into the jugular vein. If you tighten a piece of binder twine around the neck you will see the jugular vein swell up, it is toward the lower side of the neck in a slight furrow. Finding the vein is one thing, for the inexperienced, getting the needle in the right spot is another. If blood runs freely out the needle you have hit the spot.

4. Into the streak canal of the teat (*intra-mammary*) Always work the injected material up into the gland cistern by massaging upwards.

With the administration of all drugs and the treatment of wounds, strict hygiene practices must be adhered to. Clean the place where you plan to inject with cotton wool dipped in metho. Boil needles or use throw-away syringes and needles (see p. 60 calf rearing). Have animals well restrained.

Some health problems and diseases you may encounter

Listed below are only those not covered elsewhere in this book. I emphasise that the list is far from exhaustive and that any comprehensive coverage of cattle diseases is beyond the scope of this book.

Footrot or foot abscess. The animal will be limping with a foot hot and swollen. Responds well to sulpha drugs S/C and streptopen ® given I/M.

Three day sickness or Ephemeral fever. An insect borne virus which is endemic in Northern Australia and occasionally causes problems in S.E. Australia usually in late summer and autumn. The cow may be unable to rise for 2-3 days and appears to be affected in the joints. Lifelong immunity usually results from a single infection. The disease is sudden in onset. Cows run a high temperature, shiver, slobber and discharge mucous from eyes and nose. Most animals fully recover. While animals are down supply with plenty of water and try to get them into a shady spot.

Hypocalcaema (milk fever) A disease which may occur within a couple of days of calving is hypocalcaema (sometimes called milk fever or parturient paresis). This is not an infectious disease but is caused by the stress of calving which results in a lowering of the blood calcium level. It begins with a slight depression and wobbling of the hind quarters and rapidly progresses to a very severe depression with the cow down on the ground unable to lift her head and the body temperature below normal. When first seen in the morning it may even appear to be dead and is indeed close to death. An immediate injection of a suitable calcium compound (there are a number available) into the blood stream will, however, quickly revive it usually with no subsequent damage. Less severe cases may be treated partly or wholly by subcutaneous injection which is less risky than I/V injection which, if administered too rapidly, can cause sudden death from heart failure.

Grass tetany is due to magnesium deficiency and usually occurs in animals grazing young oats but is so widespread in N.Z. that magnesium oxide (causmag) is often added automatically to water troughs. The animal appears highly excited agitated and shaky. Calves may die of Mg deficiency within a matter of minutes of the onset of symptoms.

Ketosis or **acetonaemia** The cow is off colour due to the incomplete metabolism of fats, carbohydrates or protein. This produces an accumulation of acetone or dimethyl ketone in the body resulting in a characteristic acetone (sweet, sickly) smell in her breath, milk etc.

The disease can have a number of causes, such as an excess of grain in the ration, or it can result from anorexia (the cow is not eating). It is probably wise to contact a vet or experienced farmer if you suspect ketosis.

Mastitis is the most important disease of commercial dairy cattle in developed countries. It is an infection of the udder caused usually by staphylococcal or streptococcal bacteria. The presence of the bacteria on the teat does not always lead to infection. There is usually also some weakening of the cow's resistance due to stress, or injury to the udder or teat inner surfaces.

Faulty milking technique or machine operation

greatly increases the risk of mastitis incidence. Hand milking carried out hygienically and correctly is less liable to damage the teat orifice and sensitive interior linings than even the best designed and adjusted milking machines yet available.

Dipping each teat in a proper teat dip after milking greatly reduces the risk of infection and is routine on many commercial dairies.

House cows purchased as heifers and kept in virtual isolation from other cattle may not even come in contact with the bacteria. However, there is always the chance that older cows have been infected and could well carry the disease in a chronic form. Feeling the udder for lumps and left to right imbalance in quarters is one way of checking. If a cow is dry when inspected, chronically infected quarters tend to be larger than the adjacent front or rear quarter. In lactating cows chronically infected quarters will usually be smaller but clinically infected quarters may be swollen and larger than normal.

When mastitis produces symptoms such as a distinct change in the colour and consistency of milk, swelling, inflammation or tenderness in varying degrees in the udder and occasionally raised body temperature and shivering, it is known as *clinical* mastitis. Peracute cases of clinical mastitis may be fatal or lead to the sloughing off of various amounts of udder tissue (gangrene).

However, the more common form of mastitis is the *sub-clinical* infection. The milk appears normal but production in the infected quarter is reduced and there may be irregular flare-ups of clinical mastitis in the quarter. Milk from cows with sub-clinical mastitis does not usually pose a significant health threat to humans.

Abnormal milk may not be due to mastitis. During the first couple of days after calving there may occasionally be signs of blood in the milk from one or more quarters. This is not particularly serious and occurs as small pink or red clots which sink to the bottom of the milk. It is caused by the rupture of some of the small vessels in the udder following the onset of lactation. It is often associated with udder oedema or swelling.

Milk from cows with clinical mastitis is abnormal to some degree. In mild cases there are occasional clots of milk which look like custard. The hand milker is often aware of the passage of these clots through the teat orifice as the milk squirts out in staccato fashion. In more severe cases the milk may vary from a thin red or brown secretion to a discharge of the consistency and colour of thick custard. This may infect one or more quarters at the same time.

When treating clinical mastitis it is necessary to strip out all the infected discharge, preferably into a container of antiseptic to reduce the spread of infection, and then to treat the infected quarter with antibiotics in a special form for use in the udder. These are obtainable from most rural chemists and vets in the form of a small tube with an elongated nozzle which is inserted into the teat orifice after disinfecting the teat with cotton wool soaked in antiseptic. The contents are then squeezed into the udder and massaged upwards. Massage of the udder with soft soap twice daily also aids healing. The antibiotic treatment contains a blue dye which colours the milk for several days as a reminder that milk from treated quarters should not be used by humans for at least three days after the last treatment. The treatment should usually be given for three to five days depending on the severity of the infection and the drug used. There are many brands of intramammary anti-biotics available and their use is governed by the type of bacteria causing the particular infection.

The degree of recovery depends on the history of the infected quarter. If it has had a long history of repeated outbreaks the chances of cure are remote. However, early disgnosis of initial or isolated infections may lead to complete recovery.

However, the term recovery here needs some explanation. There is almost always some permanent damage done to the udder following infection and there is always the possibility that pockets of infection may remain in a sub-clinical form only to cause the development of clinical symptoms later.

Because the lactating cow continually dilutes and washes out the antibiotic placed in her udder and because such antibiotic preparations only reach infection within and not beyond the alveoli and ducts, other methods of treatment have been adopted.

These include the systemic (intra-muscular) injection of antibiotics to lactating cows, and the treatment of quarters at drying-off time with a long acting preparation (known as dry cow therapy). The latter practice is a widely adopted routine on many commercial dairies and is recommended for any quarter with a history of infection or one which has given a positive reading to a culture, cell count or rapid mastitis test (R.M.T.).

The rapid mastitis test

This test reveals the level of leucocytes (white blood cells) in the milk. These cells are used by the body to fight infection and their presence in milk is usually an indication of infection. However, freshly calved cows and ones almost dry may also show a positive

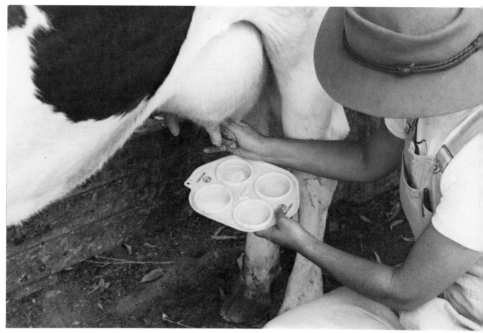

reaction when they are not actually infected. This is due to the fact that the test measures the amount of DNA (a substance found in all living cells) in the milk. The reagent or indicator used is not specific for germfighting body cells and it also reacts to dead udder and milk secretory cells being discharged in the milk.

The test consists of taking a squirt or two of milk from each quarter prior to milking and adding at least an equal amount of the reagent from a squeeze bottle.

Once the reagent is added it is mixed with the milk by swirling about 10 times.

Reading the test

The results are usually graded 0, 1, 2, 3 depending on the degree of thickness of the slime formed.
0 no evidence of slime — healthy quarter
1 thin slime can be seen as the paddle is tipped from side to side — suspect quarter
2 thick slime — when swirled it tends to move into the centre of the cup — positive quarter
3 very thick slime or soft clot, raised in the centre and retracted from the edges of the cup (looks like a raw egg) — strongly positive quarter.

While R.M.T. kits are available commercially, the house cow owner can get by with a light coloured flat bottomed cup or enamel mug and a commercially available detergent. As. many washing up detergents change their formula frequently (or purport to) the following oil company ones are listed as examples:

The type of paddle used in commercial dairies for R.M.T. There is an arrow on the paddle which is pointed to the front of the cow and thus making identification of the quarters easy. The paddle is tipped to allow all but about 2 ml of the milk to drain away.

	Dilution rate of product with water necessary to form correct reagent strength
Shell Teepol	1:3
Golden Fleece Superwash	1:3
B.P. Comprox	2:3
Mobil Detergent	2:3

Some people find the addition of some food-colouring dye aids in providing contrast for easy detection. Always use clean water for making up the reagent.

In conclusion, as emphasised under the heading of milking routine, hygiene at milking is very important in mastitis control and is the biggest single factor with house cow owners.

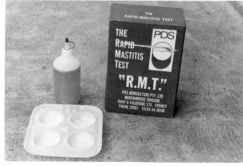

A commercially available R.M.T. kit which sells for just under $7.00 retail.

11 Owning Stud Cattle

Some house cow owners enjoy registering their cattle with a breed society and take a real interest in pedigrees and may wish to enter their animals in a class at the local show. As mentioned in the first chapter, this activity could be of particular interest to the children and teenagers.

Getting into the act

Registered animals are not necessarily more expensive to purchase than unregistered ones but there is a much greater price range. Animals with characteristics considered by the breed classifiers and show judges to be highly desirable may prove very expensive. As house cow owners and hobbyists are by and large unable to officially production record their cows, they do not usually purchase the top animals in a particular breed. Occasionally the committed hobbyist may loan cows to a commercial herd for a lactation or two in order to have them officially recorded. Herd recording is a means of measuring the milk production of cows over a lactation by metering and sampling daily production once a month and forwarding results to a central herd recording organisation. With official recording the task of identifying the cow and measuring the sample is carried out by an accredited employee of the organisation. Without production figures to back 'good looks', stud dairy cattle will never engender the confidence that brings in top money.

However, there are few commercial stud breeders on dairy farms that make much money from their stud activities anyway. They enjoy stud breeding as a hobby which gives added interest to the routine of milking cows twice a day, they have the hope of breeding a 'perfect' cow and it isn't a particularly expensive hobby.

It costs around $20-$30 per annum to be a full member of the average breed society with associate membership as cheap as $5. Registration of each animal is only a few dollars.

Show societies vary as to their insistence that entries be actually registered. Some require registration certificates to be produced.

It is usually possible to have ownership of registered cattle transferred by the breed society to your name without actually joining the society but you cannot register stock or transfer them to others without becoming a member.

When you purchase a stud animal the owner arranges to have the animal transferred to your name through the breed society. The registration certificate is forwarded to you via the society. There is usually space for such transfers to be recorded on the certificate itself much the same as the format of a motor car registration certificate.

To register animals, first join the society and you will receive a book of entry forms. Simply fill these in and send away with the fees without undue delay after the calf is a few days old.

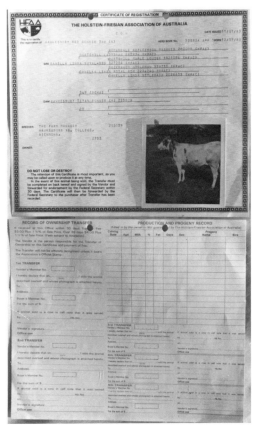

Registration certificate of the Holstein-Friesian Association. (Only Friesians require calves to be photographed for identification purposes. Such pictures are taken from the left side).

Exhibiting dairy cattle at agricultural shows

It is presumed that the original motivation behind showing was that of getting an independent and unbiased expert to class cows at a district function which had the added attraction of competition. Breeders would obviously wish to present their cattle in the best possible light and hence the development of the skills and traditions of show preparation.

Some may question the scientific value or the morality of showing, as not all desirable breed characteristics are related to milking and some practices in preparation may be looked upon as cheating. An example of the latter would be the balancing of an otherwise unbalanced udder by varying the times of milking of individual quarters or the amounts of milk removed from them on the day or night prior to parading. Whilst such a practice or the administration of a stimulant to heighten animation and alertness may occur on occasions, the average dairy farmer and show person abides by the conventions and judges are also people of experience and not so easy to deceive.

All that grooming, rugging, parading, clipping and washing are meant to do is to display to the best advantage, the natural attributes of the cow. A good milk producer tends to have a shiny coat, and be alert, lean and have good body capacity. All the preparation in the world can't make a silk purse out of a sow's ear. The quality of the animal is by far the most important ingredient in success, but it is easiest to see her potential when she is shown at the right age, stage of lactation, udder pressure, body condition, and can be paraded and stood up to allow her to look her best. Some people find a lot of pleasure in making an animal look its best.

Preparing cattle for showing

a) **Date of birth and calving date** Some cattle breeders plan their showing strategy years ahead and mate cows likely to breed show winners to bulls which may correct any faults in their offspring. They also plan the date of such matings so that the birth date of the offspring will allow maximum age advantage in the early years of its show life. For example, many show societies state in their regulations that all cattle turn one year older as at the first of January (much as horses do on the first of August).

This planning of the date of calf drop is, however, often thrown out by the desirability of having entries in the 'in-milk' classes not being too 'stale' or too long since calving. A milking cow usually looks her best either just before or in the weeks immediately following calving.

To aid planning, obtain a past copy of the show catalogue and note the classes, prize money and regulations. The chief dairy cattle steward or a friend who is an experienced show person can also prove invaluable in giving advice regarding particular local conditions and regulations.

Some major shows at times have regulations which include the necessity for entries to be production recorded or from production recorded parents.

b) **Selecting the entry** It is wise to assess the likely standard of entries and the number in any particular class before proceeding to prepare your animal.

Commercial farmers often arrange matings (both birth and calving dates) well in excess of the numbers required and then select the best representatives of the breed from among those available. These are then prepared according to a pre-determined plan. A further withdrawal of entries may occur right up to the point where the cows walk into the ring. Some capital city shows require entries to be made months in advance, while the average country show receives entries on the ground on the day of judging.

c) **Criteria for selection** The first chapter dealt in general terms with visual appraisal and selection of the dairy cow. However, there are certain distinctive breed features which every society considers are fundamental characteristics of the breed and essential to its uniqueness. Some of these, such as the dish in the face of a Jersey, or some particular colour or marking do not have any correlation at all with milk production. Large size likewise while desired by some societies, has no scientific justification; on the contrary, the average sized cow is usually the more efficient.

Text books on breeds and advice from experienced breeders will help you here. However, there is no learning experience to match actual showing which is fundamental to our gaining confidence.

There is no better way to learn these aspects of cow selection than by joining a society of your interest and/or attending shows and watching and listening to the judge, exhibitors and bystanders. Learning to judge, select and show cattle is an art and skill that no one is too old to learn from or to learn.

d) **Teaching the animal to lead on a halter** Chapter 2 dealt with the aspects of taming and socialisation. It is essential to achieve a fair degree of tameness before any attempt can be made at breaking-in to lead. Submitting to a halter is a learned response which does not come naturally to cows and must be taught, some cows learning much more quickly and completely than others.

With any conditioning process, a determined and systematic effort over a few days is much more likely to succeed than the haphazard and lackadaisical one.

There are a number of approaches, depending on the size and temperament of the animal. Obviously a full grown cow is more than a match for the average human in any tug-o-war, whereas a calf under 6 months may meet its match at the same game. So starting young is the best, where and when possible. Tying the animal up by a halter and hand feeding and watering till it settles down is one method. This will at least overcome the almost automatic tendency to pull back. The next step will depend on the size and strength of the animal compared to that of the leader, the size of yard and quality of fencing in the vicinity and the temperament of the animal.

Leading may be attempted where the animal is unlikely to clear out or cause injury to itself and the handler. The leader should utilise the point of balance principle standing at that point on the left of the beast and stepping forward or backward from that point whenever the beast either strains ahead or props. It is quite possible for one person to teach a co-operative animal to lead without additional help. However, the use of a fenceline and a person behind to either slap the beast or twist its tail at the same time as pressure is put on the halter will greatly speed up the learning process.

The aim in leading is to walk *with* the animal, not drag it. The correct position is shown below.

Three types of halter: leather, hackamore and cotton.

The halter should fit correctly. With sisal halters it may be necessary to dismantle and put a knot or two in the nose band which, unlike the remainder of the halter, does not adjust to the size of the cow's head.

Do not leave a sisal halter on a cow during wet weather as this type of halter shrinks considerably when wet and some knots can be very hard to undo.

If the animals are large an alternative method of getting them to submit to the halter is to get them to teach each other to lead.

Strong and neat fitting gear is essential with a chain of only about 16 cms long between the pair (see illustration). Swap around the collar and halter on the pair, twice a day for the first two or three days to prevent soreness and then once every day or so for a few more days. Finish off leading individually in a confined space. This method is excellent in getting larger animals to submit to pressure on the halter. 'Hackamore' halters are best

The position of the person here is about 15 cms in front of the 'point of balance'. The halter of the above cow is not fitted correctly. The strap behind the ears should have been lengthened to bring the nose band about 8 cms further down the nose.

The heifer on the left has a collar and having the mechanical advantage, it pulls the heifer with the halter about.

for this method because they are adjustable, relax immediately the pressure is off and have a chain under the chin to give heightened effect.

However, heavy use of a hackamore on an animal unaccustomed to leading can cause pain under the jaw.

The use of electric prodders and dogs or tractors and motor vehicles to drag animals along is for extreme cases only and then they may not work. Some animals simply fall over when pulled too hard and would wear their sides away on the ground as they are dragged along rather than get up. Others just bolt ahead and run the rope over or under the rear wheels, bumper bar, etc.

e) **Parading and standing up** In the show ring the cow that parades slowly with head high and eyes alert is far more likely to look attractive and catch the judge's eye than the one plodding along well behind its leader with its head hanging down.

Many showmen now walk backwards in order to have a more complete view of the judge, their beast and the procession of other cattle parading in the class.

Parading cattle. Showman walking backwards to gain a more complete view of the judge, his own beast and other cattle in the same class.

If it is hard to raise the head of a cow, tie it with its head up in the air for a few hours each day and then lead it and stand it with head up immediately afterward.

The Illawarra cow shown on page 10 is an excellent example of how to stand a cow up as the judge takes a closer look. Her halter is correctly fitted. A cow can be stood in such a way as to minimise rather than accentuate any faults of conformation.

f) **Correct body condition** The cow just referred to is also an excellent example of correct body condition for an in-milk class. She is sharp and angular yet exhibits strength and body capacity. Apart from the natural attributes of the cow this appearance is enhanced by optimum length of time since calving, correct feeding and a large quantity of milk being produced. A cow calving in fairly fat condition will take longer to milk-off that fat and hence will not reach the required angularity for a few weeks. So the showperson will try to feed each individual cow according to her present body condition and expected date of calving in relation to judging day. The closer the two days are together the leaner or finer the body condition required at calving.

Cows shown dry, and heifers can be somewhat heavier in condition but no dairy animal should be shown fat.

g) **Grooming and rugging** High producing cows have short shiny coats because of their high metabolic rate and necessity to reflect external heat and dissipate heat generated by their body metabolism. Very high producers do their best in still-air conditions well below freezing point.

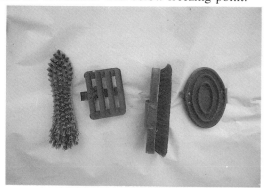

Some brushes used in grooming.

Rugging simply makes the cow hotter thus encouraging the shedding of longer hair. It also tends, by its movement over the body, to work loose hair out of the coat and thus rugging greatly reduces time required in grooming. Rugs also keep the animal clean, especially when being housed or transported. They may be purchased via

A cow wearing a correctly fitted rug.

advertisements in farm magazines or home-made out of jute bags, canvass and old blankets.

Grooming works any excess hair to the surface, stimulates circulation to the skin and adds sheen to the coat.

A rubber curry comb is used to work out the excess hair, then a stiff body (dandy) brush is used to work out the scurf and a fine body brush to lay down the hair and add lustre. In all cases brush in the direction of the hair. A scrubbing brush can substitute for a dandy brush.

Grooming can take up to 15-20 minutes a day for a large cow.

h) **Hoof trimming** This is not usually necessary but some cows just seem to walk on their heels and their toes grow faster than they are worn down. Once this starts it tends to get worse as the foot tilts further back. Trimming is the only solution. In very bad cases the toes are so long that not all the excess horn can be removed at one time without pain, bleeding and the risk of infection, so a second trim may be necessary some months later.

Hoof trimming is best done following a day or two of rain so that the horn is soft and easy to cut — much the same as fingernails are after a shower.

Hoof badly in need of trimming

Hoof after trimming.

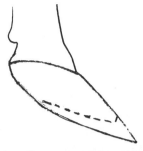

The sketch indicates approximately where to trim. Take the horn from the rim of the hoof only, coming up from the bottom. It is better to proceed slowly than take off too much and have the cow bleed profusely. Should this occur then wash the hoof clean, tie up with a leg rope and cauterise with a soldering iron.

Pictured above are two types of hoof trimmers. However, most types of tightly adjusted and sharp household secaturs will do the job. A rasp can be used to smooth out the rough corners.

Some cows will submit to hoof trimming whilst standing up, especially if they get used to it. The occasional old cow will let you do it while she's lying down, but often the easiest way to do a satisfactory job safely and quickly is to throw the cow with a long rope. Cows with very overgrown hooves may need a number of trimmings as you can't take too much off at any one time.

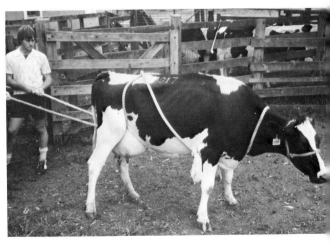

One method of placing a rope for throwing a cow.

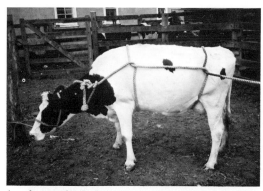

Another method of placing a rope for throwing a cow.

Three different types of clippers.

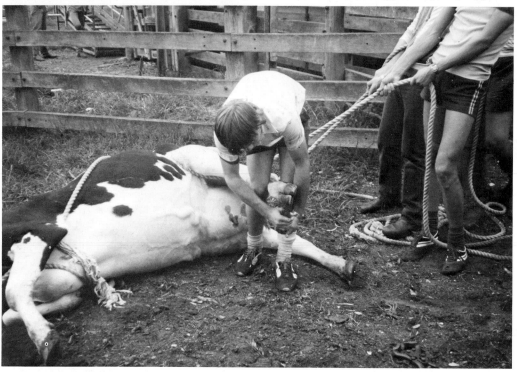

Trimming the hoofs of a cow that has been thrown. It is often quite possible to lift up or simply tie up one hoof at a time and trim it, or place it on a solid plank and cut off the excess horn with a sharp chisel.

i) **Clipping** The head, ears, udder and underline, tail above the switch, tail head and any humps on topline are the usual areas clipped. Clipping is usually "against the grain" of the hair except when blending a clipped and a non-clipped area or just skimming the longer hairs off the neck and shoulders. Clipping is usually done a few days before show day using coarse blades.

Hand clippers can be used but the job is very slow and laborious. Some friend may lend you a set of electric clippers if it is just the odd occasion. Dip the clippers blades in distillate or kerosene every now and again during use and oil regularly.

j) **Washing** This is usually done when the animal is first rugged so that the rug does not get too dirty, and again the day before the show.

Wet the cow, then pour (depending on her size) about 150 mls of stock shampoo along the top line and work it down and around the body with scrubbing brushes. Then add some more shampoo till the animal is lathered all over, scrubbing the hocks, feet and lathering the tail switch especially well.

Wash out the shampoo with a hose and squeegee out water. Some show people then rug the cow wet,

Before and after — the same cow on the same day before and after washing and clipping.

using blankets, others wait till she has dried right out. The former method is especially useful if the cow is not washed early in the day or the weather is very cold; it also usually produces a greater lustre.

k) **Tail plaiting** After washing, some exhibitors plait the tail switch into two or three braids whilst the hair is still wet. It is combed out when dry at the show. Some breeders use bags to keep the plaited tails clean.

l) **'Bagging-up' the udder** Some breeders such as those showing Guernseys and Friesians do not usually parade their cows with as much milk in their vessels as Jerseys and Illawarras. Some udders carry a lot of milk to advantage, whilst a tight udder in others will only accentuate faults such as splayed teats or lack of cleavage. Therefore the showperson needs to give thought to when is the best time to last milk the cow before showing, whether to milk all quarters right out, or whether to milk some quarters, say three hours after others. If you are uncertain, experiment some days before show day and note the times when the cow is milked and what she looks like when the interval to judging time has elapsed. It is wise not to touch the udder at all once the last milk-out has occurred otherwise the cow may 'let-down' which tends to send milk to the bottom of the udder and make the rear attachment look narrower and looser. In other words let-down can change the udder shape because once the milk is down it does not go back up.

m) **Show Day countdown** List all gear required well in advance; go over the list and check that you load all of it. Such a list may include grooming equipment, scrubbing brush, bucket, shampoo, shovel and/or pitchfork (depending on bedding arrangements), straw for bedding, hay, show schedule, attractive halters, clean rags (to wipe away manure spots), etc.

Many show or breed societies request or require leaders to wear dust coats.

Arrange transport well in advance and additional help. Allow ample time to load cattle — things can go wrong.

Arrive at the show early so that cattle have time to settle down. Check with chief dairy cattle steward regarding any queries you might have such as where to tie-up your cows and where to make entries into various classes.

Watch the show schedule closely so that you do not miss your class and make yourself known to the steward(s) for your breed.

Note what other exhibitors do and when parading keep your eye on the judge.

Beecher Arlinda Ellen. This Holstein-Friesian holds the world record for milk production for a lactation.

2 Judging Competitions

The Official Judge should determine the Official Placing and establish by number the margin of difference between each of the three pairs. These numbers represent the penalties for switching the top (T) middle (M), and bottom (B) pairs and as such form the Basis of Grading.

The total of all three penalties cannot exceed 15. If they total 15, the middle number cannot be larger than 5. If they total 14, the middle number cannot be larger than 8.

A contestant makes six decisions when he ranks four animals. This score card penalizes a contestant the amount of the margin between the two animals involved in each incorrect decision.

Computing slide for scoring judging contests.

There has been a resurgence of interest in judging competitions over the past few years in Australia mainly due to the introduction from North America of a computing slide for scoring the contests.

This slide has revolutionised marking because:

a) It automatically blocks off all irrelevant combinations once set, and makes marking very quick and easy.

b) It allows for the differential or degree of difference between any two cows in the order of placement to be considered.

In the illustration above the judge has placed the cows 3 4 2 1 with 'cuts' of 4, 2 and 1. His card would be filled out as follows:

The Royal Agricultural Society of N.S.W.
ROYAL EASTER SHOW
19.....

DAIRY CATTLE JUDGING COMPETITION
(Non Oral Classes)

Competitor's No. or Name: ......*Judge*..

Age: ....

Breed & Class: ....

Placings:	1st 3	2nd 4	3rd 2	4th 1	Total
Cuts	4	2	1		

An explanation of 'cuts' — the basis of grading or marking

Before the slide was adopted, contestants lost the same number of marks for any switch of placings from those of the judge no matter how close or far apart the judge considered the cows to be.

In the picture of the slide above, at the bottom of the column, is the 'basis of grading'. The judge believed there was a 4 marks difference between the top two cows, 2 marks between the middle pair but only 1 mark between the bottom pair. In other words he felt there was a fairly clear winner in the class and contestants should have placed her first. However, there was, on the other hand very little difference between the cows placed third and fourth and anyone who had them around the other way should not be penalised very much — only one mark.

With the fast and fair marking that this slide allows, more classes can be judged allowing a much more accurate and just assessment of the contestants. Using 'cuts' also allows for the possibility of the organisers and/or the judge to reduce the effect of a particular class by reducing the total of the 'cuts'. These can run from a total as low as 3 to as high as 15.

This means that if one class of four cows is particularly difficult, confusing, or of doubtful value as a measure of a person's judging ability, then the total cuts given will be low and even if the contestant has the cows completely reversed, he or she will still gain considerable marks, whereas the contestant who does the same thing when the total of cuts are high, as in a class that is easy to place, will lose practically all marks.

DAIRY CATTLE JUDGING COMPETITION
(Oral Classes)

Competitor's No. or Name: *John Smith*

Team: ....

Breed & Class: *Holstein-Friesian Cow*

	1st	2nd	3rd	4th	Max. Marks	Marks Gained
Competitor's Placing	3	2	1	4	50	
Oral						
Accuracy of observation					25	
Comparative rather than descriptive statements					15	
Style, ability and time					10	
TOTAL					100	

REASON CARD FOR COMPETITORS NOTES

Class: *Holstein-Friesian*

Placing: *3214*

3 over 2 (Top Pair) stronger front & rear attatchments better body capacity	Concede **2 over 3** (Grant) slightly more dairy charact
2 over 1 (Middle) (Pair) more dairyness throughout. Stronger & cleaner thighs. Legs, stronger loin.	Concede **1 over 2** more attractive in head & neck
1 over 4 (Bottom) (Pair) very little difference stronger in legs	Concede **4 over 1** greater capacity in udder. Fault 4 Sickle Hocks.

An oral judging card. The portion on the left is handed in.

The importance of judging contests

a) They provide an important opportunity for training and for potential show judges to seek excellence.

b) They help in raising the level of motivation of young people who may pursue agriculture as a career.

c) They help in sharpening decision making skills.

d) They provide a social venue for young people to meet and exchange ideas.

e) Give helpful feedback on public speaking.

f) Most important, the contestants learn about cows — how they are built and how they work, and how to relate form to function.

How judging contests are run

Any number of classes may be run. In North America it is usually six in national contests. At present in Australia it is generally three, with 10 minutes allowed per class. The cards are marked or graded within minutes and the 6-8 contestants with the highest totals then give oral reasons on a predetermined class. For that class (one of the three they have just done) the contestants used a different card. This card is bigger and they tear it in half and keep the half on which are written notes which can help them with their two minute explanation of how they placed the four cows and why.

Above is an oral judging card. The portion on the left is handed in.

The following information on how to give oral reasons comes from the Purdue University Dairy Judging Manual which was based on one developed at Cornell University.

Judging competition at Menangle NSW.

Reasons in International Judging Competitions

Giving oral reasons

Judging contests are frequently won or lost on the scores given the contestant for his oral and written reasons. Reason scores are based upon:

- Accuracy and completeness
- Organisation (order, identification, pairs, timing)
- Terminology (proper, comparative terms)
- Presentation (poise, forceful, confident, smooth delivery)

You are allowed two minutes to give your reasons. The judge will be listening for certain important facts on each pair of cows. Have your reasons organised and present them in a clear and concise manner. Presenting a set of reasons in one minute is much better than two minutes of talking with only a few facts.

The main objective of your reasons is to inform the judge that you saw and compared each animal in the class. To do this well, you must know dairy cattle and have a complete vocabulary of dairy cow terms.

The most common error made by boys and girls who are learning to give reasons is to *describe the animals* instead of *comparing them*. An example of this is "I placed the number 1 cow first because she was large, straight over the topline, and had a level rump. Her udder was attached strongly, both fore and rear." This does not tell the judge why you placed number 1 over number 2. You should say, "I placed number 1 over number 2 because she was a larger cow which was straighter over the topline, and had a *more nearly level* rump. Her udder attachment was long*er* and stro*nger* in the rear." By using *er* words such as *stronger* instead of strong, *straighter* instead of straight, you will be making direct comparisons that will tell the judge why you placed the class as you did. Do *not* use the words *good* or *better* because they do not indicate difference.

When giving reasons, you must be able to visualise completely each animal in the class. Do not memorise your reasons, but "see" the animals as you talk about them. If you memorise your reasons and forget one sentence, you may be unable to get started again. You *are not permitted* to use notes when giving reasons.

When practising your reasons before giving them to the judge, try to find a quiet place in the room and give your reasons aloud. You will gain confidence by hearing your voice, and you may notice that some phrases are not clear.

Give your reasons in an authoritative and enthusiastic manner. Show some conviction that you have placed the animals correctly. Do not discuss your placing with anyone before giving your reasons. To learn that several others have placed the class differently will only make you nervous and uncertain of yourself.

The first impression you make on the judge is very important. Adjust your clothing and remove your hat before approaching him, then make your appearance in a dignified and composed manner.

Give the judge your complete attention and expect the same from him. Look him straight in the eyes. If this bothers you, look at his forehead or just over his head, and he may not be the wiser. Do not begin giving your reasons until the judge has asked you to start.

Stand straight, on both feet, and do not move or look about. Place your hands at your sides or behind you; do not wring your hands or twirl a pencil.

Put some depth in your voice. Breathe deeply and let the words roll out strong and clear. Emphasise the most important points by making your voice deeper and louder. Do not try to blast the judge out of his chair, but adjust your volume to the conditions.

Order or organisation of reasons

The most accepted order of giving reasons is as follows:

Top pair: Compare — grant or concede — rarely fault the second place cow.

Middle pair: Compare — grant — sometimes fault third place cow.

Bottom pair: Compare — rarely grant' — fault last place cow.

Start your reasons with: "I placed this class of Holstein aged cows 4-1-3-2." Give your placing with firmness so the judge can write it down or get the placing in mind. Then follow with: "I placed 4 first and over 1 because she was ." Give the most important reasons first. Remember to use comparative terms. "I grant (or admit) that 1 has straighter rear legs than 4."

Continue in the same manner with reasons for placing 1 over 3 and 3 over 2. Discuss the last placed cow as follows: "I fault 2 and place her last because she . . ."

You may end your reasons with: "For these reasons, I have placed this class 4-1-3-2."

The judge may ask several questions to clear up a point, or to determine whether or not you saw a certain point. If you know the answer, make an immediate, concise reply. Do not ramble on to other points because you may confuse your answer. If you do not know the answer, reply, "I do not know or I do not remember." Do not guess.

Sample set of written reasons

Comparative phrases

Many boys and girls can place a class of dairy cattle easily and also recognise the differences between the cows; however, they frequently have difficulty in telling the judge their reasons for placing one cow over another.

If you can recognise the differences between cows and familiarise yourself with the phrases listed below, you will be on the road to giving accurate, comparative reasons. When you are giving your reasons, you will not have to think how to make a certain comparison. This will be easy to do after you have the basic phrases well in mind. With experience you will develop phrases of your own. These are only a few to help you get started.

Breed character

She shows more Holstein type and breed character, especially about the head and neck.

She has a stronger jaw and is wider at the muzzle.

Dairy character

She is far more desirable in dairy character, being sharper and leaner over the withers, hips, and pins.

She shows more openness of rib, and is flatter and cleaner in the thighs.

She is longer and cleaner in the neck, and it blends more smoothly with the shoulders.

Body capacity

She has greater body capacity, being longer bodied and deeper in both fore and rear ribs.

She has greater strength of heart, being wider on the chest floor, has greater depth and spring of fore rib, and is deeper in the rear flank.

General appearance

She is a larger cow with more size and scale.

She has a straighter and stronger topline, and is more nearly level from hips to pins.

She is smoother and wider over the rump, higher at the thurls and is neater over the tail head.

She is tighter, stronger, and smoother at the point of shoulder, and blends in more fully and smoothly through the crops.

She shows more style and balance, especially on the move.

She stands on a straighter set of rear legs as viewed from both the side and rear.

She is flatter and cleaner in her bone and has more substance of bone.

Udder

She has a larger udder that is attached more strongly, and more smoothly in the fore, and attached higher and wider in the rear.

Her teats are more squarely placed, hang more nearly plumb, and are of more desirable size.

Her udder appears to show more quality as indicated by more veining.

Comparative and descriptive terms

Breed type and general appearance
Positive terms

Has a more feminine head
Has more style and balance
Has more dairy quality
Is smoother throughout
Neck blends more smoothly with the shoulders
Has more strength and substance
Shows more style when on the move
Stands more correctly on her rear legs
Is straighter on her rear legs as viewed from the rear
Stands more squarely on her legs
Has shorter and stronger pasterns
Is deeper in the heel
Is flatter and cleaner in the bone
Has a wider and more nearly level rump
Is wider and more nearly level from hips to pins
Is higher at the thurls
Is wider at the pins
Is a more upstanding cow

Is stronger over the back and loin
Is longer bodied and more upstanding
Is longer and more open in type
Is stronger through the front
Is tighter through the shoulders
Is smoother at the point of shoulders
Is fuller through the crops
Is stronger through the heart

Negative terms

Is rough over the rump
Is sloping over the rump
Is low at the pins
Is high at the tail head
Lacks over-all balance
Lacks in breed type about the head
Lacks quality and refinement
Is too refined and lacks strength, constitution, and vigor
Lacks smoothness
Lacks strength of top
Is weak in the back
Is weak over the loin
Short bodied cow and too close to the ground
Lacks width of rump and is too narrow at pins
Lacks size and scale
Lacks condition and bloom
Is too coarse about the head
Is plain in the head
Has wry face (twist from the eyes to the muzzle)
Has wry tail to the right, left
Is crooked in the rear legs
Is sickle hocked
Is cow hocked
Stands too close at the hocks
Is too long in the pasterns
Is weak in the pasterns
Toes out in front
Is shallow in the heel
Does not walk freely

Dairy character
Positive terms

Has a more milky appearance
Is more dairy like
Is sharper over the withers
Is more angular throughout
Is more open ribbed
Is longer in the neck
Is thinner and cleaner in the thighs
Is cleaner at the throat
Is flatter and more open of rib
Is cleaner over the top with a more defined vertebrae
Has a neater and more refined tail head

Negative terms
Is heavy over the shoulders
Is short and thick in the neck
Is throaty
Is thick in the thighs
Is parchy at the hips and pins
Is close ribbed
Is too thick and meaty over the rump
Is rather heavy and coarse at the tail head

Body capacity
Positive terms

Has more body capacity by ——
Is deeper bodied
Is longer bodied
Shows more stretch
Has greater depth of body
Is deeper and more opened ribbed
Is deeper in the flank
Has more spring of fore rib
Is fuller in the crops
Is deeper through the heart
Is wider on the chest floor
Is fuller in the heart girth
Is wider down the top

Negative terms

Is pinched at the heart
Lacks depth of heart
Cuts in behind the shoulders
Is narrow chested
Lacks spring of fore rib
Is shallow bodied
Is too flat in the fore rib
Is cut-up in the flank

Mammary system
Positive terms

Has a more evenly balance udder
Has more balance of rear quarters
Udder is more strongly attached both fore and rear

Has more strongly attached fore udder
Has stronger rear udder attachment
Has higher and wider rear udder attachment
Has stronger and smoother fore udder attachment
Stronger centre support in the udder
Udder shows more desirable halving and quartering
Udder shows more balance from front to rear
Is more firmly attached in the front
Carries her udder closer to the body
Has more width between rear teats
Teats are more evenly spaced
Teats hang more nearly plumb
Is more desirable in size and shape of teat
Has a more balanced udder
Has more uniformly placed teats
Has more quality of udder
Carries out fuller in the rear quarters
Has more balance of the fore quarters

Negative terms

Weak in fore and rear attachments
Loose in front udder attachment
Is too short in the fore udder
Is too low and narrow in rear attachment
Has unbalanced udder
Is light in the right rear quarter
Is too deep in the udder
Udder shows too much halving and quartering
Teats strut
Has funnel-shaped teats
Rear teats are too close
Is bulgy in the fore quarters

A milking shed

Some points to avoid in giving oral reasons

• Do not start giving your reasons until told to do so by the judge.
• Do not slouch.
• Do not chew gum or have anything in your mouth.
• Do not move around, jump, shift from one foot to another.
• Do not give "canned" (memorised) reasons.
• Do not describe the animals — use comparative terms.
• Do not use terms "better," "good," or "nice".
• Do not call the judge "Mr Judge".
• Do not "guess" when answering questions.
• Do not *over* criticise the animals — their owner may be listening to the reasons.
• Do not use "bag" in describing the udder of a cow.
• Do not use tortuous milk veins and milk wells.

Housing

In much of Australia this is seen more as a concession to the milker than the cow. Milking in the open in very windy conditions can result in dust, etc landing in the milking bucket and the milk being blown "off course" and splattering everywhere. Rain results in water running off the milker's sleeves and hat and off the cow's back (plus dirt and hair) and landing in the bucket.

Any sort of 'lean-to' on the side of a shed can provide the shelter when required. However, it is often more satisfactory to build a proper little milking shed, with a bail feeder, feed storage, shelving, hooks for hanging halters, and an eyelet low down on the wall for attaching a leg rope.

The shed pictured opposite has all those features.

The base of the feed trough should be at least 9 in above ground level to prevent the cow stretching and dancing about trying to get all the feed. When feeding concentrate only in the bails a 20 litre plastic drum with a bit off one side is ideal (see page 5). If, however, you wish to feed hay, a 200 litre (44 gallon) drum split vertically (as in picture) is ideal, however, it may cause the cow to 'chase' her concentrates around the big trough during milking and hence fidget and dance about.

It is usually necessary to pave an enclosed area (bricks, concrete slabs as in picture, or concrete) but hard packed clay can work. It is important that the cow stands up hill with front feet up to 3" higher than back feet. This is especially so for earth floors. With this slope, when the cow urinates or defaecates, the stuff falls further behind the cow, hits at an oblique angle and sprays away from the milker and drains away easily. The slope also makes the cow put her legs back for balance purposes and that makes the back teats of the udder easier to get at and with more weight on the back legs kicking is also more difficult.

Glossary

Anthelmintic a medication administered to animals to treat internal parasites.

Bail a mechanism for restraining the head of a beast. It is usually a wooden or metal bar that pivots at the base and swings across a vertical plane (see photo on page 15).

Bobby a young calf, usually male and usually sold for slaughter.

Crush a restraining device at the end of the race which holds cattle while various management practices are carried out.

Cutting out segregating an animal from a group.

Dry cow a cow that is not lactating. Usually refers to a pregnant cow within two or three months of calving.

Elastrator a special type of pliers that expands a small and very strong rubber band so that it may be placed over the testes or say the tail of a lamb. The rubber band cuts off the blood supply and then causes the part to die and slough off.

Electrolyte replacer a mixture of chemicals, mostly salts in solution in water which is used in treating calfhood diarrhoea (scours). It is usually administered orally but a special formulation can be injected.

Fore milk the first milk removed from a cow at a particular milking. It is considerably lower in butterfat percentage than the cow's average secretion.

Free-martin a female calf, one of twins of which the other is a male. Almost 90% of free-martins are infertile.

Heart girth the girth or circumference of the cow immediately behind the shoulder.

Join mate a cow to a bull.

Let-down a neuro-hormonal reflex in the cow that causes milk pressure in the udder to rise making is much easier to remove.

Milked-out a lactating udder which has all the available milk removed.

Milk replacer a commercially formulated mixture containing mainly spray dried skim milk powder and homogenised tallow plus added vitamins and minerals.

Race a narrow passageway usually just under 700 mm wide (I.D.) for holding cattle.

Stale cow one which has not calved recently and whose current lactation has continued for many months.

Strip to extract milk from the cow by sliding the thumb and forefinger down either side of the teat with sufficient speed and pressure to cause milk to be expressed. This technique is often adopted to get the remaining milk from a cow after conventional hand milking has removed most of it.

Strippings the last milk removed from a cow at a particular milking. It is considerably higher in butterfat percentage than the cow's average secretion (more creamy).

Further reading

Further specific reading on stud cattle, showing and judging might include:

American Dairy Cattle by E. P. Prentice, Harper Bros, New York and London.

The Dairyman (Aust. Stud Dairy Cattle Monthly), Eltham, Victoria.

Dairy Cattle Judging Techniques by G. W. Trimberger, Prentice-Hall.

Hoards Dairyman Judging Guide. Fort Atkinson, Wisconsin.

Preparing Dairy Cattle for Show. Hoards Dairyman.

There are many other books on cow husbandry and hobby farming. In the former category are books such as:

The Herdsman's Book by Russell and Williams, Farming Press, Ipswich, England.

A Veterinary Guide for Animal Owners, Rodale Press.

and in the latter category:

The Healthy House Cow, by Marja Fitzgerald, Earth Garden Magazine, Melbourne.

Index